G

D1255878

MIRACLE OF THE AIR WAVES
A History of Radio

Today radio spans the globe. Tomorrow it almost certainly will beam messages from outer space. Yet just a little over sixty years ago it was hailed as a near-miracle when a freighter in the North Atlantic picked up the sound of a human voice broadcast from a scant two hundred miles away. Among the colorful personalities who paved the way for the radio revolution of the 1920's were Marconi, deForest, Fessenden and Armstrong. This is the dramatic story of radio's amazing growth from a scientific toy to an immense industry—the extraordinary scientific pioneers who fostered it, and the tremendous transformations it has worked on the 20th-century world.

BOOKS BY EDWARD A. HERRON

ALASKA'S RAILROAD BUILDER
Mike Heney

CONQUEROR OF MOUNT McKINLEY
Hudson Stuck

FIRST SCIENTIST OF ALASKA
William Healey Dall

MIRACLE OF THE AIR WAVES
A History of Radio

MIRACLE OF
THE AIR WAVES

A History of Radio

by Edward A. Herron

illustrated with photographs

JULIAN MESSNER Ⓜ **NEW YORK**

Published simultaneously in the United States and Canada by
Julian Messner, a division of Simon & Schuster, Inc.,
1 West 39 Street, New York, N.Y. 10018. All rights reserved.

Printed in the United States of America

SBN 671-32079-3 cloth trade
671-32081-5 MCE

Library of Congress Catalog Card No. 69-13043

MIRACLE OF THE AIR WAVES
A History of Radio

chapter ——— *one*

On Christmas Eve, 1906, the freighter *West Selene* was wallowing along in the North Atlantic, bound from New York to Rotterdam. Two hundred miles to the west was the bleak, winter-stricken shore of New England.

Henry Jacobson, the wireless operator aboard the *West Selene,* adjusted his earphones and scratched absentmindedly on a pad of paper while he listened for the snapping *dit-daws* of a message coming through the air. There was little traffic. The world was engrossed in holiday preparations. Once or twice Jacobson, bored, twisted about and looked sleepily at the jumbled arrangement of coils and wires that crowded most of the wireless quarters. He wished for the end of his twelve-hour shift.

Suddenly he flicked the pencil to one side and clapped his hands so sharply to the phones pressing on his ears that he winced. He sucked in his breath in amazement. Without turning, his hands still pressing the phones tight to his ears, he blew into the speaking tube leading up to the bridge and screamed, "Captain! Captain! I'm hearing a man's voice and a violin! Captain!"

At the dawn of that winter day, the rays of the new sun lightened the sand shore of the mainland at Brant Rock, in Massachusetts, outlining a lonely tower that thrust upward from the beach 30 miles south of Boston.

The wind whipped at the sand, and the sand scratched at the metal tower. For a few moments the faint hiss and scratch of the sand on metal was the only apparent sound.

December 24, 1906, was a momentous day. On that day the world started to talk.

Radio, the miraculous dispatch of the human voice, did not have a well-ordered birth. It lurched into view, then slid back into the scientific laboratories for fully fifteen years before it finally escaped and started on its explosive career, bringing entertainment and instant communications to the entire world.

It is difficult to grasp the idea that such a marvel would have such a grudging acceptance.

In the second half of the twentieth century, startling engineering and scientific accomplishments are commonplace. In just ten years, for example, inventions have brought man almost to the magic moment when he will land on the moon. And with the moon almost in our grasp, our scientists are embarked on a similar conquest of the ocean depths. There is no doubt, no hesitation. Using the combined talents of great engineering organizations, the lower oceans will be conquered and their secrets uncovered.

The birth of radio, in sharp contrast, was dragged out for nearly two and a quarter centuries. A Dutch mathematician, Christian Huygens, started the blind groping in 1678 when he suggested that light might be the result of rapid vibrations producing some kind of an invisible wave.

The host of other laboratory experimenters who followed in the 18th century became more and more convinced of the existence of a mysterious wavelike motion. Ben Franklin knew there was "something" out there when he risked his life with the daring kite experiment in 1750, pulling bolts of electricity from the thundering skies.

Michael Faraday, in 1832, moved a step closer when he suggested that electricity in magnetism is sent through space in some sort of vibration.

Other scientists backed up his theory with laboratory demonstrations in which they were able to flick magnetized needles into motion, even though the needles were separated from the electrically charged wires by 200 feet of clear space.

But it was a German scientist, Heinrich Rudolph Hertz, who opened the door. In his laboratory he finally created, detected and measured the elusive electromagnetic waves.

Using crude laboratory equipment, Hertz proved that electric waves could be sent out at will. He set up two loops of wire some distance apart. In one loop, he left a break across which a spark of electricity would leap whenever the current was thrown on. When that happened, a similar leap took place almost simultaneously across the gap in the loop 200 feet away. The waves, he said, traveled at the speed of light; they could fill the universe.

It was Hertz who gave a name to the "something" that had puzzled scientists for centuries. There was a special medium which he called ether, the odorless, tasteless, invisible substance that fills all the available space in the universe—ether, the invisible medium which transmitted the waves of light and carried the all-invading electrical waves.

In the scientific laboratories where work with the magnetic waves was being conducted, the excitement was building up to a climax. So sure were scientists about the path to be followed that a British physicist, William Crookes, could write, in 1892:

Here is unfolded to us a new and astonishing world, one which . . . should contain possibilities of transmitting and receiving intelligence.

Rays of light will not pierce through a wall, nor . . . through a London fog. But the electrical vibrations of a yard or more in wave length . . . will easily pierce such mediums which to them will be transparent. Here, then, is revealed the bewildering possibility of telegraphy without wires, posts, cables or any of our present costly appliances.

This is no mere dream of a visionary philosopher. All the requisites needed to bring it within grasp of daily life are well within the possibilities of discovery, and are so reasonable and so clearly in the path of researches which are now being actively prosecuted in every capital of Europe that we may any day expect to hear that they have emerged from the realms of speculation to those of sober fact.

What remains to be discovered is firstly, a simpler and more certain means of generating electrical waves of any desired wave length. Secondly, more delicate receivers which will respond to wave lengths between certain defined limits and be silent to all others. Thirdly, means of darting the sheaf of rays in any desired direction, whether by lenses or reflectors. . . .

It is interesting to speculate what might have happened to the still unborn radio industry if some one man had entered the world stage in 1892 and electrified the scientific world with a challenge and pledged support in achieving world-wide radio—just as President John Kennedy electrified the engineering world of America and sent it resolutely onward to the task of conquering the moon.

There were great engineering companies who could have listened and had the capability of achieving the goal. Western Union, Postal Telegraph, the American Telephone and Telegraph Company, General Electric and Westinghouse were already giants in their fields in 1892. They were the logical ones to leap into action and open up new pathways in voice communications—but none of them did.

Perhaps if the German Hertz had lived, instead of dying in 1894 at the age of thirty-seven, he would have perfected his crude laboratory instruments, changing the aimless stuttering into an orderly procession of dots and dashes. Or if Hertz had not carried his pioneer work forward, it could have been any one of a dozen other scientists working in lonely, isolated laboratories throughout the world.

The man who read Crookes's prediction and swung into action was a most unlikely candidate. He was not a highly trained engineer, but a young Italian boy, sheltered in the quiet seclusion of a wealthy family.

They made an odd pair, an old, sightless man and a young boy, sitting in the shade of an olive tree. On a wooden box before them was a practice telegraph key. The young boy's fingers balanced delicately above the brass rod.

"Flex the wrist gently, so. Now again—and again —and again. That is better."

Guglielmo Marconi listened intently to the insistent tones of the blind man. In response to his urging, his fingers moved almost imperceptibly, and from the slight motion of the key came a torrent of cricket noises, the dots and dashes of the Morse telegraph code.

The old man nodded in satisfaction, then hoisted himself slowly to his feet and reached for his staff. "Very good. You send better now than I did when I was the operator at the railroad station. Very good. I will be here again tomorrow, God willing."

Guglielmo Marconi was born April 25, 1874, in Bologna, high on the northern portion of the Italian boot. His mother was Irish, a student who had met the wealthy elder Marconi, married him and borne him two sons, Guglielmo and another boy nine years older, Alfonso.

Far removed from the turmoil that occasionally gripped the land, the family lived in quiet seclusion. Tutors came to the family villa and taught the boys. There were protracted journeys to England and long stays in southern Italy, where the sun was kinder during the winter months.

Guglielmo was 14 years old when he became deeply interested in science. From childhood he had always been in the company of university professors, friends of his father. Some of his tutors were excellent in their chosen scientific fields.

When he learned of the early struggles and eventual triumph of the American, Samuel Morse, in establishing the first telegraph systems, he determined to learn all that was known about telegraphy. His first venture was to seek out the old blind man, a former telegraph operator, and learn from him the Morse code. His easy familiarity with the *dit-daws* of the interrupted signals hurrying over telegraph wires was to have a significant bearing on some of his future electrical experiments.

The manner in which Marconi could concentrate on a project was almost ferocious in its intensity. When he read of Franklin's experiment with the kite, nothing would satisfy him until he, too, had rigged up an experiment akin to that of the Philadelphia philosopher. He built an antenna on the roof of his father's home, deliberately inviting the intrusion of electric bolts. In the midst of a violent storm, while bolts of electricity ripped in the black, high-banked clouds, his antenna "sucked in" static electricity, and the bell he had hooked into the circuit rang.

He was completely absorbed in the wonders of electrical science. Nothing distracted him. When he was persuaded by his brother to go along on a mountain climbing expedition, he went almost reluctantly, impatient to get back to his experiments. His knapsack, instead of being crammed with the necessities of high-altitude life, was filled with scientific journals. During a rest period he pulled out one of the magazines and there he read the predictions of William Crookes, the Englishman. The secrets of wireless telegraphy were on the verge of discovery!

Instead of wild elation, the young Marconi, just 18 years old, was seized by a fit of depression. As he hurried down from the mountain, he was convinced that someone else would announce the achievement he wished so ardently for himself.

His first task when he rushed back to the two-room laboratory he had furnished atop his father's house was to rebuild, piece by piece, the apparatus Hertz had used in his experiments that proved the existence of electromagnetic waves.

He reproduced the Hertz experiment without difficulty, with a maximum distance of 30 feet. Then he examined Hertz's ideas critically, substituting materials, and with each change he got an improvement. He moved his apparatus out into the family vegetable gardens and bridged 100-yard gaps. When he devised an aerial system, the world's first, he spreadeagled his signals for a phenomenal distance of one mile.

Hertz, in his original experiments, had sent out big, stuttering charges of electricity that gushed across the intervening gaps in the hoops almost like the flood of water spurting from a hose.

Marconi chanced upon a better way, one which he immediately patented. It was a simple move, but a vital one. He introduced the Morse telegraph key into the circuit.

Immediately he had the signals under close, intimate control. When he opened and closed the key, he sent out the identical *dit-daw* signals that the Morse system was sending over the confining wires.

That was in 1896. Marconi was 22 years old.

With his father's encouragement, he first offered his invention to the Italian Government. Ignored, he went to England for vivid demonstrations of the new system of wireless telegraphy. Many important scientists watched the demonstrations as he beamed signals over distances as much as nine miles apart. How many of them were chagrined that a youngster could have devised what they themselves had been searching for for generations? Some of them spoke unkindly of his efforts, saying in effect, "After all, what is it that Marconi did? He simply improved on already existing ideas."

Marconi took the experiments out of the laboratory and turned them into a practical device that would influence the course of nations. His effort was to result in the formation of a huge corporation that would dominate international wireless communications for two decades.

The whole world watched as he confidently prepared to bridge the Atlantic Ocean with his *dit-daw* signals.

The navies of the major powers competed for the right to install his new signaling system. Merchant ships plying lonely courses, far removed from land, began to use the Marconi system of wireless telegraphy. The world was started on a new era of communications.

Marconi tended his inventions, raced in fast cars, sailed in fine yachts and claimed he had no control over the oddities that crept into the company that bore his name, the British Marconi Company.

Very rightfully Marconi was acclaimed as one of the great inventors of all time. For most of the world he was the object of the greatest respect; for a few men he was a constant source of irritation.

One of those irritated was an American, Lee deForest.

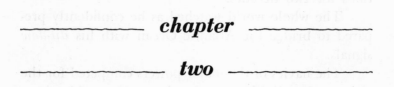

chapter

two

Lee deForest had the "jumpies." That was his own candid evaluation when he went up the slippery gang-plank of the night boat that made the peaceful south-bound journey from Milwaukee to a new job in Chicago. Under deForest's arm was a cardboard box full of wireless parts; in his mind was an odd mixture of insolence and self-assurance that both captivated and repelled his audience, the gullible stock investors of America.

In the first decade of this century, countless investors were attracted to this clear-eyed, nervously alert deForest, promoter, indefatigable inventor, scribbler of patent applications.

America, in the early part of the century, hungered for heroes, and Americans turned readily to inventors emerging from the new and exciting scientific world. Marconi was already acclaimed, but he was under a handicap, a double foreigner, an Italian "masquerading" (this was untrue) as a British subject. DeForest was true-blue American. He was well educated, and was actually repelled by uneducated people. Lee deForest was destined to go from rags to riches. He received his Ph.D. from Yale in 1899. He was not the most brilliant young man to get his doctorate, but he was the hungriest.

He went through Yale undergraduate and graduate schools on a starvation diet that featured 20-cent meals. He made no bones about his hunger while he pursued research in the new excitement of the time, Hertzian electromagnetic waves. As much as possible he followed the activities of young Marconi, who had shifted his wireless experiments from Italy to England.

At age 26, deForest started to work for a living. He eased into Chicago, apprehensive, desperate and still hungry. He was forced to accept a job at $10 a week in the dynamo factory of the Western Electric Company. He hated his work and he disliked the rough, uneducated people who surrounded him in the shop. His sole recreation was a once-a-week 25-cent seat at the Chicago opera.

He had need for little else. The consuming interest of his life was wireless telegraphy. His doctoral thesis had explored the mysterious ether waves that spread in every direction, unseen. In Chicago he turned from his thesis to actual hardware, and piece by piece he built a crude wireless set, clearing a space in the jumbled heap of books, old magazines and soiled laundry atop the dresser in his attic bedroom. He worked practically day and night on improving the apparatus. He worked in his lonely bedroom, and he worked at Western Electric, absentmindedly shoving aside the company's work to concentrate on his own.

When Western Electric objected, he accepted a job as an assistant to another inventor, also interested in wireless telegraphy, and with a better salary, $15 a week. Very shortly he decided the inventor's wireless telegraph system would never work, and once again he shoved his employer's work aside and concentrated on his own. He was caught and fired.

He starved gently, with an air of detachment, for his mind was not on his hunger, but on his wireless telegraphy apparatus. His first device was an automatic detector that made possible the use of the telephone for wireless signals.

He rigged odd-looking antennas on Chicago rooftops, sending and receiving Morse code signals that were clearer and sharper than the Marconi efforts.

He knew his system was good. He was so engrossed in the worth of the wireless telegraph system that, for a time, he completely ignored the possibility of transmitting the human voice through space.

He gauged almost all of his hidden efforts in relation to the well-publicized efforts of Marconi. If the Italian reached out for 100 miles with his signals, deForest hungered for 200. When Marconi talked of the possibility of sending Morse code messages across the Atlantic, deForest thought of blanketing the United States with a network of telegraph stations. When Marconi hinted that he was about to come to the United States to conclude a contract for a trial of his equipment by the U. S. Navy, deForest dreamed of equipping the entire British fleet with deForest equipment. At the time of his dreaming he was still experimenting atop the dresser in his bedroom.

But not for long. He was about to break out of the attic, out of Chicago, and take on Marconi in person.

The Italian had an instinct for capturing the public fancy, and deForest studied it closely. When he read that Marconi was coming to the United States in 1901 to report by wireless the big boat races held each summer off Newport, Rhode Island, deForest decided to enter the competition—not with the boats, but with Marconi. The Italian had contracted with a major newspaper to send the race results by Morse code from his vantage point, far out in Long Island Sound, directly to a shore station, which in turn would hand the message to a newsman waiting by a telephone open to the newspaper office.

To get into competition, deForest needed $1000 immediately. He was honestly convinced that his wireless system was better than Marconi's. But he needed money to buy the necessary components, build the sending and receiving sets and then get back East, not only to the scene of the races, but also to a possible newspaper client anxious for quick service on the race results.

His friends were hesitant about loaning him the money, but he spoke convincingly of a company he was about to form, of stock that would be issued. He was persuasive, and the money came in. He rounded up the necessary equipment, went back East hurriedly, interested a major newspaper and headed out to sea for a good position in the midst of the huge white sails.

The waters were choppy and the race exciting, and deForest "chattered" away happily, his expert fingers tapping lightly on the telegraph key. He had visions of the strong *dit-daw* signals that were flashing boldly across space to the waiting receiver on shore. News by wireless! It was exciting.

Unfortunately Marconi was doing the identical electronic feat, on the same wave length. It was as though two express trains were crowding into a long tunnel on a single track. The two competitors blanked each other completely. Nothing but a screeching mishmash reached the shore stations. There was no intent during the race to harm each other, but later the same type of interference became commonplace between rival wireless networks.

The information got ashore, but with no help from the wireless sets. Hand signals were wigwagged from the watching boats, and the information was telephoned to the newspapers. The newspapers had spent considerable money in the "news by wireless" feat, and for weeks had been whetting the public curiosity about the astounding feats to be accomplished by this new scientific marvel. They did not bother to tell their readers about the breakdown in communications. The headlines still screamed, "NEWS BY WIRELESS." It was false, but deForest did not protest. Neither did Marconi.

DeForest even became fascinated by the newspaper clippings, though he knew the reporting was false. He gathered them carefully, arranged them in neat rows and then, with his homemade brochure, descended on Wall Street bankers, seeking funds for expansion.

If Wall Street stayed clear, the public did not. The publicity about the yacht race brought in the first hesitant investors. DeForest sold stock in his new American Wireless Telegraph Company. With the first funds he rented a factory in Jersey City, New Jersey, and started building wireless apparatus.

The first important customer was the United States Army, which asked for a test installation at some Army forts along the East Coast. The United States Navy was always a strong supporter of research in communications. The Navy had tried the Marconi sets, and one devised by a German scientist. Now they asked deForest for demonstrations.

The Jersey City factory kept busy. But there was always a need for fresh money for expansion efforts. More stock sales were necessary. Because he recognized his limited talents in that area, deForest called in several colorful promoters who said, in effect, "You build the wireless; we'll sell the stock."

For once deForest was not hungry. He paid himself a salary of $30 a week. Good luck waited for him every time he turned around. His name and his wireless sets were respected and in demand. He was alert for the likeliest spots that would bring additional acclaim. When the Sino-Russian War was imminent, he dispatched two of his best men to a vantage point on the China mainland and relayed war information across the country to the steamers waiting to race from China ports to San Francisco. As the Russian fleet lumbered around the world toward its tragic encounter with the Japanese, wireless sets of the world's leading inventors were in active use, serving the fleet commanders well and focusing a spotlight on the new technique. DeForest's equipment operated exception-ally well.

DeForest swept from one exciting success to another: His wireless sets were aboard the express banana boats of the United Fruit Company, whose sleek vessels plied between the United States and the banana plantations of the Central American countries. The company urgently needed dependable wireless communications between its vessels and various ports as they sped the highly perishable cargo north to the United States.

DeForest not only had his equipment aboard the vessels, but also landed a contract to build land wire-less stations for the company, establishing a link be-tween the main shipping points in Costa Rica and Panama. The experience the young inventor gained in protecting his delicate equipment from the destruc-tive heat and the metal-consuming corrosion of the tropics served as a springboard to big U. S. Navy con-tracts for major wireless stations at Pensacola and Key West in Florida, at Guantánamo Bay in Cuba, in Panama and in Puerto Rico.

The Navy contract gave deForest particular satis-faction, because he won it in hot competition against the Marconi Company. He tossed restlessly in his sleep in the hot tropical nights, eager to add to the string of triumphs. He was, he thought, on his way to the wealth that had eluded him. Unfortunately, he had lost control of his own destiny.

DeForest, despite the flood of contracts, was always desperately in need of money. He turned to several stock brokers, who readily agreed to float the necessary funds. DeForest was suddenly at the head of a company capitalized for $3,000,000. While the inventor busied himself trying to complete the new wireless installations, his stock partners busied themselves with promoting new, sky-high schemes, all of them woven around the magic of the deForest name.

The inventor claimed he never realized what was going on behind his back. But there were many who accused the young man of knowing exactly what was going on, and actively assisting in the planning.

One of the plans was for the erection of a vast network of wireless land stations that would, in one bold move, rival the two existing giants, Western Union and Postal Telegraph. Literature pushed by the promoters spoke of ridiculously low rates that would be charged for telegrams. To add impetus to the sales, the returns from the first wave of stock sold to the public were used to erect nearly one hundred steel transmitting towers in all corners of the United States.

The towers themselves, though many of them never sent a message, acted in turn as a powerful impetus to the sale of stock. Stick to deForest, investors were told, and it will be as though you had invested in the first days of the Bell Telephone Company.

In justice to deForest, it must be pointed out that for much of the time during this frantic promotion, he was out of the country, working for endless hours through day and night, trying to complete the contract for the installation of the U.S. wireless stations in the Caribbean area.

He was tanned, healthy, alert and fantastically busy as he moved about in the warm tropical sun, supervising the wireless installation.

During the construction period deForest continued, as he had ceaselessly since the day he had left Yale, to improve his wireless apparatus. He was no occasional inventor. He made changes almost daily in the transmitting and receiving equipment.

When he returned to the United States after arduous labors, he discovered he had been too busy. The flashy promoters he had hired to push the stock of the American Wireless Telegraph Company had overextended themselves. Before deForest had unpacked his bags, his "partners" told him that the company was in serious financial trouble. DeForest was horrified. He knew that hundreds of investors would lose their savings. He sensed the outcry that would be made against anyone associated with the company. Abruptly he resigned from the American Wireless Telegraph Company, accepting as severance a cash payment of $500 and retaining sole rights to one of his earlier patents. He watched the company collapse. Once again he was out in the street.

Darker hours were still ahead for deForest. He had been engaged in a running court battle with another inventor, Professor Reginald Fessenden, and suddenly he lost the battle. By accident or design deForest had slipped across the boundary of patent rights and freely used the idea of Fessenden's important liquid barreter in wireless stations and sets that he, deForest, had built. To the watching public, the Fessenden-deForest suit was just another of the many battles going on in the wireless industry. Almost simultaneously deForest, Fessenden, Marconi and other inventors were suing and being sued. They were at times plaintiffs, at other times defendants. The nation's leading scientists were paraded through court rooms, labeled as highly amusing pirates.

But the loss by deForest in the Fessenden lawsuit was serious. It was 1906; he was seven years out of Yale; he was 33 years old and almost penniless.

Professor Reginald Fessenden. The name should have been remembered with regret by deForest for the rest of his life. Actually, it was Fessenden who burst open the door that permitted deForest to leap to the heights of acclaim.

chapter

three

The last, intense preparations for an event that would influence the world for generations to come started on a December night in 1906 on a train leaving Schenectady, New York, heading eastward for the grinding haul over the Berkshire hills to Boston.

Twisting uneasily on the hard seat of the train, 40-year-old Reginald Fessenden almost reached out physically for the forgetfulness of sleep. But he had ample reason for his wakefulness. He was a fastidious man with a trim beard and mustache. The soot and grime of travel that had blackened the green velour of the railroad coach seat annoyed him. He was a big man, well over six feet tall, with broad shoulders and an expanse of waist that just came short of being fat. His long legs were cramped in the narrow space between the seats even before the journey had started.

But it was his angry frustration that kept him tossing about in the darkened coach.

The train began to pick up speed. From time to time the yellow glare of sleeping towns flared on the dusty window. Albany was a maze of lights and subdued night activity with the train crews running alongside, speaking in muffled tones over the hiss of the engine. But beyond Albany, after the Hudson river had been crossed and the train started penetrating the Berkshire hills of Massachusetts, the night lights dimmed and almost vanished. In the dragging predawn hours Pittsfield, North Blanford and Springfield each flared and died in darkness, the mournful greeting from the train's whistle menacing the sleep that Fessenden was so earnestly wooing.

He moved uneasily, looking across the aisle, up to the water cooler, then back over his shoulder to the battering noise pouring from the open door leading to the baggage car.

He thought of the big wooden crate which carried his name, Professor Reginald Fessenden. *Professor* Fessenden? Or had he left that title behind when he quit the University of Pittsburgh six years earlier? That had been a momentous burning of bridges. From that day forward, he, like his idol and mentor, Thomas A. Edison, would follow but one star. He would be an inventor, Fessenden had vowed, and, almost arrogantly, he repeated, he would be a *rich* inventor. The talents God had given him—and he was convinced he had them in greater measure than most men—would pay off handsomely. No one was going to steal from him the fruits of his labors.

He grinned and rubbed the heel of his palm on the dirt-grimed seat. His efforts had been profitless so far. Fessenden had his own dream, and part of it was hidden in the big wooden crate in the baggage car. Fessenden was convinced that he could superimpose the human voice on a continuous flow of electrical current without wires, just as Alexander Bell had placed voice on an electrical current flowing through wires.

There were two keys that would help him. One was the new alternator, guarded in the baggage car, built for Fessenden by Dr. Ernst Alexanderson of the General Electric plant in Schenectady. That would give him the smooth flow of high frequency current he would need. The other was a Fessenden invention, the liquid barreter, an amazing advance in wireless reception that he kept locked tight, so he thought, in patented secrecy.

All previous wireless signal detection systems had depended on a unit called the coherer, which was able only to distinguish the dots and dashes of the Morse code. Fessenden's liquid barreter, which looked like an electric light bulb, was far more sensitive.

The United States Navy had bought sets using the

Fessenden barreter; the fleet of the United Fruit Company used them. The Army came and requested an installation for use in Alaska. Mexico sent representatives to investigate. The Postal Telegraph System erected stations on Chesapeake Bay, all of them using the Fessenden set and its magical liquid barreter.

It was encouraging, but Fessenden had bigger dreams. He knew positively that the liquid barreter was so sensitive his system would be able to accomplish what no other system had been able to do—pick up the sound of the human voice superimposed on invisible electromagnetic currents.

He knew he could do it. He had done it before. He had heard the faint whisper of a human voice, almost drowned in the crackling and boom of static.

He sweated nervously, thinking of the ridiculous failure of the previous September. One small, broken wire in the alternator, the result of a careless, stupid . . . . He bit his lips to stop the flow of adjectives. A railroad employee had thoughtlessly dropped the crate containing the vital piece of machinery, and the jar had snapped a wire, one wire of thousands. The entire experiment had been wasted. His patient financial backers tried to hide their disappointment.

Now the repaired alternator was on the train, in the express baggage compartment. This time, he would see that no one erred, that the machine was delivered safely. If it failed. . . . He shrugged and kicked his foot against the seat in front of him. There was a stifled yelp, then a head lifted and peered over the seat, "Would you please—"

"I beg your pardon, ma'am." Fessenden pulled his long legs back. If he failed? Who would know except the few faithful people who had stayed with him through years of disappointment?

He drummed nervously with his stiffened forefinger on the train seat. The dots and dashes of wireless telegraphy had been crossing open space for several years in a remarkable demonstration of a new kind of communication.

The goal was to get a signal strong enough and clear enough to be heard across the Atlantic, tying North America to the British Isles.

Fessenden, with the enthusiastic backing of financial partners, with whom he had formed the National Electric Signalling Company, was convinced he could vastly improve the methods of transmitting and receiving long-distance signals, and that he could win the race to link the two continents.

There was more than glory involved. To the victor would go important contracts, perhaps exclusive contracts, from the various governments involved to make permanent wireless installations. To the victor would go control of a new world industry.

Up to this point the concentration was on the *dit-daw* signals of wireless telegraphy. In 1906 only one man, Reginald Fessenden, had in mind an even more startling feat—hurling the human voice across the oceans.

The idea was not bizarre. The trick was to get rid of the thunderclaps of noise that raced side by side with the *dit-daw* signals. A hundred improvements and refinements were necessary; entirely new approaches to the transmission of signals were needed. It was in the area of improvements and refinements that Fessenden exhibited his greatest confidence.

He was justified in his faith. He and a small group of other inventors were about to perform near-miracles in ingenuity. Unfortunately, some of the miracles overlapped, and the seeds of dissension were being planted at the moments of greatest triumph.

Fessenden's experiments were expensive. In Scotland and at Brant Rock, Massachusetts, he built enormous towers 400 feet high. The towers were cylindrical steel tubes, almost like smoke stacks, wide enough for a man to crawl upward inside. They were firmly anchored by guy wires to the ground.

Enormous towers, high-frequency signals, the liquid barreter for sensitivity—this was Fessenden's attack, and it brought him success, first in wireless telegraphy, then in his secret dream, voice transmission in space.

On January 10, 1905, for the first time in the history of the world, two-way transatlantic Morse code telegrams were sent and received from Brant Rock, Massachusetts, over to the receiving station in Scotland.

For three days there was intense jubilation while signals *dit-daw*ed triumphantly back and forth across the ocean. But at the very height of the triumph, before the world was even aware of what was happening, the entire system went dead, a victim of one of the most prolonged electrical disturbances ever known on the North Atlantic. The system came to life briefly, sputtered and went dead again, permanently locked in steel-laced silence. C724000 CO. SCHOOLS

For months Fessenden labored to bring the system back to life. Marconi's efforts, further north, were likewise unsuccessful, but that was little solace for Fessenden. To add to his despair, a storm ripped across Scotland and blew down the twin tower at Machrihanish. It was totally destroyed. It would cost too much to rebuild. Millions of dollars had already been spent.

Any lesser man would have quit, but Fessenden continued the struggle. This time he brought boldly to the fore his primary intention, to broadcast the human voice. He ordered from the General Electric Company in Schenectady a new, extremely high-frequency alternator, the heart of his system. It was that unit that had been dropped, and his experiment had failed.

Now he was returning with the repaired alternator.

His mood was so depressing that he jerked viciously on the stiff window shade, as though trying to make the darkened interior of the train coach even darker, and so bring on his much-needed sleep. The lady in the front rose again in protest, but this time Fessenden glared at her belligerently. His temper was notoriously short. The lady sensed the explosive atmosphere and turned away, fingers pressed to her ears.

Fessenden stared ahead into the darkness. The train rattled and banged forward. The hours passed, and the sun began to rise as the wheels raced along the track.

When the train pulled into the station at Boston, Fessenden rose stiffly to his feet, tipped his derby hat to the lady, who glared at his indifferent politeness, and then walked back to the baggage car. He watched carefully as the crate was transferred across the platform to another train for the last few miles south to Brant Rock.

Once there he started with his assistants to install the repaired alternator. Late on Christmas Eve he turned wearily to Fred Givens. "Let's try it." His hand reached for a switch. Then he lifted a clumsy microphone, a narrow box wrapped in asbestos, protection against the hot wires that were inside.

He gave a short speech, speaking in a restrained voice, almost as though he were doubting that his words were carrying beyond the narrow room. When he had told his unseen audience what he intended to do, he motioned to his wife, Helen, and she turned on the phonograph. The music of Handel's "Largo" shrilled tinnily in the room. Fessenden moved the microphone closer, and then pulled it away, the movements betraying his nervousness. Gradually, as the music swelled, he relaxed. He sat in a straight-backed chair, and closed his eyes, listening.

When the music died away, he lifted his violin, and without hesitation played the carol "O! Holy Night."

When that was finished there was an awkward silence while he stared helplessly into the microphone. His wife leaned forward and held out a bible, her finger pointing to a text. In a strong, vibrant voice, Fessenden declaimed, "Glory to God in the Highest, and on Earth peace to men of goodwill."

He started to pick up the violin again, changed his mind and blurted out, "Merry Christmas to all!" He started to switch off the instruments; then he came back hurriedly to the microphone. "Will those who have heard these words and music please write to R. A. Fessenden at Brant Rock, Massachusetts. We will speak to you again on New Year's Eve."

Afterwards Fessenden walked down to the beach.

What if he had spoken into empty space? What if no one had heard?

But they had.

The wireless operator on the *West Selene* was only one of many operators on ships within a few hundred miles of Brant Rock who had heard the amazing assortment of speeches, violin solos, biblical quotations, phonograph records and the plea that they write to Fessenden. Some of the vessels were heading into Boston, and within two days scrawled postcards came from the wireless operators. "We heard your music Christmas Eve. We could make out most of the words."

Reginald A. Fessenden was granted U. S. Patent No. 706,747—the first issued for voice transmission by electromagnetic waves.

One week later, on a blustery New Year's Eve that ushered in the year 1907, Fessenden "broadcast" again, and this time his voice was heard at a U. S. Navy station in Puerto Rico.

The striking demonstration of Fessenden's theory, that it was possible to send the human voice over long distances without the aid of wires, so elated his financial backers that they poured even more money into the venture. A sum of two million dollars ultimately was invested in the jointly owned National Electric Signalling Company.

The inventor expected the world to react in the same way, but the world paid little attention.

Fessenden plunged back into his experiments, and the quick result was another great invention that was to have a profound impact on voice radio communications. Fessenden's heterodyne circuit mixed a signal having a certain frequency with a high-frequency current generated by a carbon-arc oscillator. The end result was a receiver more sensitive and more static free than any ever known before.

He was ahead of his time. The heterodyne circuit was destined to gather dust for seven years, waiting for another key that had not yet been invented.

Fessenden was a devoted scientist, a tireless inventor, but he never reached the fame, or the notoriety, of some of his contemporaries. Only a few men recognized his genius, and it is unfortunate that those very men who conceded his worth were to become his bitter enemies, locked in hopeless lawsuits with the inventor.

Choleric, demanding, vain, pompous—all these adjectives can be heaped on Fessenden, but none of them can hide this inescapable fact—he was one of the great inventors of the era. The lack of recognition that was his lot came not because of the quality of his contributions, but because they were too far advanced for their time. Had contemporary inventors kept abreast of the Fessenden brilliance, it is quite likely that the introduction of commercial radio broadcasting would have been advanced a full 10 years.

He was never content with the end product that came from his laboratories; he pushed hard always to build some new equipment, something revolutionary; and he did not stop to calculate the cost, or the probable sales value of his latest invention.

But it was his stubbornness and violent reaction to opposition that finally overwhelmed him.

Reginald Fessenden was born October 6, 1866, in East Bolton, Canada, the son of a minister. The family moved to the United States, and the boy received good schooling. He early showed an interest in science, but he was equally versed in Hebrew and Greek.

He could have gone to Oxford University in England when he completed DeVeaux Military College, but by a quirk that was to influence his life, and affect the life of generations to come, he decided to go to New York City and seek work with Thomas A. Edison. Street lighting based on arc lights was in general use, but the new excitement centered on the problems of devising a smaller, more efficient method of electrical illumination for interior lighting. Edison had turned his entire attention to an activity that resulted in the first incandescent lamp, and a host of new developments were springing from that discovery.

It was that activity that attracted Fessenden. Unfortunately Edison had no opening for the young man, but with bulldog determination Fessenden went to work digging ditches in the streets of New York with the gangs of workmen who were laying cables for the rapidly expanding electric lines.

Edison walked past one of the ditches, recognized Fessenden and was so impressed with his determination that he made a spot for him in the laboratory. Within a short time Fessenden was running the Chemical Laboratory and had a direct hand in some of Edison's most important inventions: the moving pictures, the system of electric current distribution, the cathode tube and the storage battery.

Not only was Fessenden thrilled to be working closely with Edison, he was also intensely loyal. In 1889, when he was only 23 years old, he was offered $10,000 a year to enter into a partnership in a thriving young business. It was a magnificent sum, but Fessenden refused. He wanted to stay with Edison. The blow came a year later when the laboratory had to shut down and Fessenden was released. It was in those bitter days Fessenden resolved that he would put a high price on his future scientific efforts. Already he could see that the formation of companies to exploit the inventions was as important as the inventions themselves. Edison's efforts were already evolving into the future supergiant, the General Electric Company.

Fessenden went to work for another scientific laboratory, where he continued to do important work on electrical instrumentation. In his free time he carried on independent experiments with alternating currents. He read voraciously in scientific magazines and books. Only a few generations before his time men had known practically nothing about electricity. Now as the knowledge mushroomed, Fessenden became one of the few experts in the field.

His fame was acknowledged when he started an eight-year term in teaching, first as head professor of Electrical Engineering at Purdue University, then at the University of Pittsburgh.

It was during that period the name of Marconi emerged. Fessenden followed every triumph of the young Italian almost with a tinge of bitterness. He checked and rechecked on Marconi's principle of interrupting the current at random points of the signal wave in order to achieve the *dit-daw* signals. There had to be a better way.

When he was sure he had found it, he turned his back on university teaching and all commercial laboratory work. From that day forward, he vowed, he would live by his inventions.

The Weather Bureau, part of the United States Department of Agriculture, was interested in the possible use of wireless telegraph stations to cover the storm-swept Outer Banks of North Carolina. Fessenden, despite his resolve to live by inventions, accepted a position with the Bureau at a salary of $3,000 a year with $4 a day expense money. If he could convince the Weather Bureau that his system was superior to Marconi's, it would be worth the temporary inconvenience of once again living on a salary.

It was during the Weather Bureau assignment that Fessenden conceived the possibility of bringing to life another dream, the transmission of the human voice without wires.

As a demonstration project for the wireless telegraph, two wooden towers, 50 feet high, were built on Cobb Island, south of Washington, D.C. The test was to demonstrate Fessenden's improved wireless telegraphy set, using the conventional *dit-daw* signals induced by the spark-key method. When that was proved to the Weather Bureau's satisfaction, an assistant spoke into a crude microphone. On the lonely site at Cobb Island, in January, 1901, intelligible speech was transmitted for the first time across space using electromagnetic waves. The Weather Bureau officials, called down for the wireless telegraph demonstration, pressed earphones to their heads at Fessenden's invitation and listened in puzzled wonder.

The scientist himself listened in disappointment. The sound was faint and broken, distorted with harsh noises caused by the flashing spark key. It was no good, Fessenden admitted ruefully. It would never be acceptable. But there was one ray of hope. The words, despite the distortion, were intelligible. Perhaps, if he could get rid of the strident irregularity caused by the interruption of the waves by the key. . . .

The Weather Bureau authorized Fessenden to build three large experimental telegraphy stations at Cape Hatteras, at Roanoke Island and at Cape Henry in North Carolina.

For two years Fessenden was completely absorbed in the installation, the dream of voice communications completely shelved. Then, his temper flaring, he accused the Weather Bureau officials of trying to steal his patented ideas, and he quit abruptly. It was a violent display that was to recur many times. His life would be spent in short, brilliant successes, long, bitter wrangling and lawsuits.

It was after his stormy exit from the Weather Bureau that he determined to get back again into the race with Marconi, and the young American Lee deForest, whose name was cropping up more and more often in spectacular news involving wireless telegraphy. Marconi was backed by a mammoth British corporation, so big that they were establishing a branch, the American Marconi Company, in America almost as though it were a corner grocery store. Fessenden had the firm backing of two Pittsburgh businessmen. They were liberal with their funds, but they were prudent also. Already communication in space had the vague outlines of a billion-dollar endeavor.

Fessenden selected the site at Brant Rock, and built the steel tower, with an invisible link to its twin over in Scotland. These were grandiose moves befitting a man who was convinced that soon his name would outshine Marconi's. He did not rank deForest as a full-grown competitor. Fessenden was exasperating with his violent displays of temper, but his backers were patient, excusing much for his genius.

The more they spent in patience and money, the more demanding Fessenden became. He was suspicious of every paper he signed. He was convinced that his business associates were conspiring to cheat him.

His attention to his experiments slipped badly as he girded for battle against those who helped him. Abruptly the patience of his backers was exhausted. They locked the office of the National Electric Signalling Company, and Fessenden was on the street.

Legal maneuvers followed, and Fessenden felt himself persecuted, a victim of harpies. The crowning blow came when he was denied the use of his inventions.

He fought in and out of the courts, winning and losing.

There was a climactic court battle in 1912 when he won a judgment of $400,000 against his two former partners.

It was a hollow victory. The National Electric Signalling Company was forced into receivership, unable to pay the judgment.

Years passed with Fessenden clamoring for his money and far removed from his first love, voice transmission in space. Finally the defunct company was acquired by Westinghouse Electric Company. The wheels still turned slowly for the payment due Fessenden. It was finally made in 1928, only a few years before his death, and it came too late to soften his bitterness.

He never experienced the acclaim given Morse, or Edison, or Marconi. The grinding dissension probably robbed America of inventions even greater than those already credited to Fessenden.

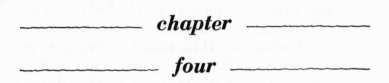

chapter

four

No one is quite so ignored as a civilian aboard a Navy ship the day the vessel is to depart on a long voyage.

Lee deForest leaned heavily on the rail and looked over to the dock at the Norfolk, Virginia Navy Ship yard. As far as he could see, the tall latticework of battleship masts rose in the air. Tied to the numerous finger docks and clustered thickly out in the calm waters that led to Hampton Roads and Chesapeake Bay and the wide waters of the Atlantic Ocean were smaller craft, colliers and auxilliaries, impatiently waiting for the big battleships to have their fill of coal and provisions. More than 14,000 seamen were anxious to be under way.

It was December 16, 1907. The Atlantic battle fleet was about to steam into history.

DeForest was too tired to be excited.

"Mr. deForest," an officer called impatiently from the top of the ladder leading below, "We're ready for you. Please hurry."

DeForest nodded. He was near the point of utter exhaustion. All of the previous day, he had ridden in a cramped railroad coach from Jersey City, New Jersey, down to Washington, D.C., then transferring to another train for the slow, bumpy ride down to Richmond and Fredericksburg. At the latter, he had changed trains again, and late in the evening, nearly 20 hours from the time he had left Jersey City, he stood shivering on the darkened platform of the station at Norfolk, Virginia. A chill breeze was whistling over the flat land, unimpeded in the rush from the nearby Atlantic Ocean. The crates of wireless sets had been unloaded from the baggage car, and deForest winced as the men set them none too gently on the horse-drawn drays that waited impatiently in the darkness.

He worked in the early dawn aboard the battleship *Kansas* and then, with that installation completed, he transferred to the battleship *Illinois*. Every minute counted.

The United States Navy was about to provide the most dramatic stage for demonstrating voice radio communication that had ever been offered any inventor. What happened to the deForest radio telephone sets in this instance could easily influence the course of voice communications for the next decade.

The battle fleet of the United States Navy was to be a solid test bed.

It was the United States Navy that had come to the rescue of Lee deForest just when the inventor thought that his entire career had ended. For six years the Navy had been experimenting cautiously with wireless sets aboard battleships, obtaining only indifferent success.

The experiments were only tolerated by the line officers. Signal flags by day and signal lights by night had been good enough for the Navy for more than 100 years. Who needed unreliable, mysterious electrical equipment for communications? That was the thought among the older officers.

But the young ones kept pressing for wide installation of the new device, pointing out its great value in helping to keep ships in formation during dense fog, of the value to an admiral of knowing exactly where every one of the ships in his fleet might be at any given moment. The enthusiasts played down the fact that wireless would also be a tight link between Navy headquarters in Washington and stubborn Navy commanders far out at sea.

The seamen who operated the wireless sets were unskilled, and fearful of the electrical mumbo jumbo. The sets were not sturdy. The continual pounding and vibration of the ships while at sea caused the frail electrical connections to snap easily.

The Navy kept trying to improve its wireless techniques. The officers at sea were still indifferent, some of them openly hostile, but the shore-based Navy was anxious for the sets to work. They wanted to use the wireless as a long finger of authority, keeping tighter control over the ships at sea.

But even the most indifferent officers of the big ships admitted that it would be nice to be able to communicate quickly with other ships in the fleet, especially in dense fog. One enterprising young officer, enamored with the success of the land telephone lines that were beginning to spread like horizontal forests over the country, received permission for a unique experiment. He trailed telephone wires from one ship to another in a long line of vessels as the fleet plowed ponderously forward through the night. On each ship a telephone was hooked into the circuit, in theory permitting the entire line of vessels to be cut into a long party line.

The telephone wires dangling in the water snarled on thrashing propellers, and the experiment was literally torn apart.

DeForest had dealt successfully with the Navy on previous occasions, and he was aware of the Fleet's increasing interest in voice radio communication between ships and between the ships and the shore stations.

Before deForest heard of Fessenden's feat of

broadcasting the human voice through space, he was at the very depths of depression.

Yet the news of Fessenden's feat erased all the depression. Irrepressible as ever, deForest raced back to his laboratory and started tinkering. He had a new goal, a new ambition, a new world to conquer. And he had a powerful weapon ready, a unique tube which he called the Audion. The tube ultimately was to become the basis of the entire radio industry. It was to gross over a billion dollars in sales. It was to make deForest wealthy.

Using the tube, he quickly devised his own system of radio telephony. He secured a contract from the Lackawanna Railroad to set up a voice communication system between a land station on the New York side and the many Lackawanna ferryboats that plied the river between Jersey City and New York. Thousands of commuters were thrilled by the new device that was now controlling their lives. DeForest made sure that suitable demonstrations were made for the newsmen who crowded aboard.

A Navy officer heard of deForest's voice radio installation on the ferryboats chuffing noisily across the Hudson River. He made the trip across the river many times, listening intently to the exchange of conversation that helped keep the big boats on schedule. DeForest was alerted, and the inventor came hurriedly to the ferry slip. He dipped into his precious funds to buy lunch for the officer, then brought him to a building in uptown New York where he was regularly broadcasting to another building about three blocks away.

DeForest casually suggested a motorboat demonstration. A week later he had the young officer afloat on Lake Erie. DeForest spoke into a microphone. "Turn right," he said in a very unnautical command. "Now turn left." The officer lifted binoculars. Two miles away a companion motorboat moved in response to deForest's terse commands.

The officer hurried back to the Navy Department in Washington. DeForest was munching on a dry sandwich when the postman came with the letter. The Navy Department requested him to demonstrate the radio telephone system aboard two battleships. DeForest, who could recognize the coming of good fortune in clouds no bigger than the palm of his hand, agreed.

The Navy asked that the apparatus be simple and steady in operation, and that it cover a range of 10 miles between ships. The most difficult requirement was that the sets be installed, within a matter of days, in the battleships *Connecticut* and *Virginia,* which were about to cruise about the ocean in simulated warfare off Cape Cod.

DeForest never had a stockpile of equipment. Whatever he fabricated he sold so quickly it had no time to gather dust, so desperate was his need for cash. He stayed all night in the small rooms that comprised the company factory. He assembled parts skillfully, but hurriedly, and the units gradually began to take shape. Toward dawn he slept for a few hours stretched on a cot in the corner of the room; then he was awake again, reaching for soldering iron and pliers, weaving the intricate circuits. For several days he and his assistants worked relentlessly, stopping only for some food brought in from a distant delicatessen store, and occasionally sleeping when utterly overcome with fatigue.

The sets were finally completed. The two battleships were scheduled to make a temporary coal stop near the Naval Station at Newport, Rhode Island. DeForest and three assistants crated the sets securely, then rode with them in the baggage cars of the train to Providence, Rhode Island. There they carefully carried the heavy equipment aboard a ferry that threaded through the islands of Narragansett Bay south to the Naval Station at Newport. As they came in sight of the white buildings almost submerged in the green of the surrounding hills, there was an imperious hooting and whistling.

The *Connecticut* and *Virginia,* bursting with full loads of coal, steamed proudly out to sea.

DeForest was not easily defeated. Within two

hours he was aboard a fast launch pursuing the two monsters that were already heaving in the open waters of Rhode Island Sound. The launch came alongside the *Connecticut,* and deForest called through cupped hands up to the bridge. There was no answer, but there was a perceptible slowing of the big ship; the plume of black coal smoke pouring from the stack faltered and sagged uncertainly as the vessel came to a halt. A ladder was lowered and deForest scampered up.

"Sorry we forgot about you, Mr. deForest. We'll lower a sling and bring your gear aboard." The tone of the officer's voice belied his concern. DeForest knew he was in unfriendly territory.

The necessary equipment was brought aboard and, with it, two of deForest's assistants who were to make the installation. DeForest himself went down the ladder. "Now, sir," he said to the officer who had greeted him cooly, "if you'll send a signal to the *Virginia* and ask them to heave to, I'll be obliged."

"I'll have to get the Admiral's permission, sir."

DeForest was already halfway down the ladder. He leaped in the launch. Over his shoulder he could see the signalman on the bridge of the *Connecticut,* flags snapping smartly as a message was sent to the companion battleship.

Despite the obstacles, and the hostility of the officers, the demonstration was a success. The Navy

asked that the voice message be heard distinctly when
the ships were at least 10 miles apart. Actually, they
were 21 miles apart in one of the tests, and deForest
was right in using the word "smashing." But Navy
historians, who had as little love for deForest as they
did for Fessenden, refer to the historic event as only a
"moderate" success.

DeForest went back ashore, determined that all
the effort he had placed into the building and installation of the two radio telephones would not be wasted.
A startling announcement from Washington helped
his determination.

President Theodore Roosevelt announced that
he was going to dispatch the Atlantic Fleet on a 16,000-
mile journey around Cape Horn to the West Coast of
the United States. The ships, on their long journey,
would have the eyes of the world focused on them.
The flotilla would be commanded by a well-known
hero, Rear Admiral Robley Evans, who had picked
up the rough-and-tough nickname "Fighting Bob."

The Washington newspapers carried a story that
Fighting Bob had been so impressed by the result of
the tests of the deForest radio equipment on the *Connecticut* and the *Virginia* that he was insisting vehe-
mently that every one of his 16 battleships in the big
Fleet be provided with the new equipment.

Navy historians say that Evans never did request

the equipment. No one knows how the story got into the papers. But the important result was this: in early November, deForest received an order from the United States Navy for 26 radio telephone transmitter and receiver sets. He had exactly 40 days in which to build the sets and install them. The Fleet was making final sailing preparation at the Norfolk Navy Base in Virginia.

It was an impossible task, but deForest nearly accomplished it. He and his assistants tossed component parts together. He made the tiresome trip from Jersey City down to Norfolk several times, superintending the installation of the equipment in the ships that were crowding the anchorage at the Navy station. He had his own crew of workmen for assistance, for the Navy was too preoccupied with its own chores to be worried about the civilians who cluttered up the decks.

DeForest and his cohorts were still struggling with the installations when they were almost unceremoniously told to leave the ships. The Fleet was about to sail.

It was nearly noon, December 16, 1907, when the flagship *Connecticut,* with Rear Admiral Evans aboard, hoisted a signal and began to move slowly out into the stream. From every ship, bands began to play. The President of the United States, who had earlier made a farewell speech lauding the effort, lifted his hand in a parting gesture.

Lee deForest stood on the edge of the pier, unnoticed. There were still four sets that had not gone aboard the ships. Already his mind was racing, planning on a future meeting with the fleet in Rio de Janeiro to complete the installation. Overhead huge plumes of smoke floated from the twenty stacks. The lead ship turned out to wider waters, easing into the Chesapeake. The vast plumes of smoke were still visible to the watchers at the base when the Fleet moved out into the gentle swells of the Atlantic Ocean. The 16 battleships lined up in precise formation in columns of four, the columns 1600 yards apart, the ships keeping a 400-yard stretch of water between the fantail of the leader and the bow of the following ship. The vast armada of battleships, colliers, and auxiliaries occupied two square miles of ocean.

How did the radio telephone equipment operate? Franklin Matthews, a correspondent for *Colliers* magazine, filed this story shortly after the great trip got under way:

> There was much interest in the ships as to how the radio telephone would work out. The system has been in operation only a few months and is largely in the experimental and almost infantile stage.
>
> All of the battleships are equipped with the apparatus and there was no doubt about it, you could talk to any ship in the fleet from one to the other, and at times the sounds of the voice were as clear as through an ordinary telephone.

At times they weren't, and there was a division of opinion among the officers as to the real value of the invention.

As in the case with the wireless telegraph, only one ship of a fleet can use the radio telephone at one time. While one ship is talking to another all the other ships must keep out of it and even the ships to which the message is being sent must keep still and not break in. The receiver must wait until the sender has got all through with what he has to say and then talk back to him.

The sending and receiving machines use part of the apparatus of the wireless telegraph outfit. If an attempt is made to use the telegraph while the telephone is in use, the telephone goes out of commission because it is absolutely drowned out. The telegraph apparatus uses so much greater power that it is like a loud voice overwhelming a soft one.

The operator at the telephone would sound a signal with some sort of a buzzer that had a wail like a lost cat in its voice, and then he would put a little megaphone into the mouthpiece and would say, sharp and clear:

Minnesota! Minnesota! This is the *Louisiana!* This is the *Louisiana!* This is the *Louisiana!* We have a message for you. Do you hear us? Do you hear us? *Minnesota! Minnesota!* This is the *Louisiana.* Go ahead! Go ahead!

Sometimes the message would fail. Sometimes the wireless, one kind or another, would be working on other ships. Sometimes the answer could come at once and the operator would write down the reply and hand it over to you.

Everyone of the electrical experts with the fleet is convinced that the wireless telephone is going to be of value. Most of them have talked with it clearly for distances of at least 20 miles. One difficulty is in keeping it tuned up because the wireless telegraph apparatus is also on board.

Some of the experts seemed to think that the wireless drops in efficiency if the radio telephone is kept keyed up to its best. All were confident that as soon as certain difficulties were overcome, difficulties no more serious, they said, than the ordinary telephone encountered in the beginning, the apparatus would be workable as readily as a telephone on land. Give it time, was the way the situation was summed up.

When deForest read the report, the first word he had of the operation of his radio telephone, he was elated. He was sure that this demonstration would be the battering ram to open the doors for unlimited use of the radio telephone.

Unfortunately Admiral Evans was old, sick and highly nervous. He was not even convinced of the worth of the regular wire sets that had been installed in the Fleet, let alone the radio telephone. The previous year he had commanded the squadron which had attempted to use the wireless in maneuvers in the Caribbean. The result had been a shambles, with signals becoming a mishmash of incoherent *dit-daws* that created indescribable confusion. The simulated "enemy" received the squadron secret messages with greater ease than the ships for whom the electronic messages were intended. In one example, the battleship *Maryland* tried for hours to call the *Colorado*. When the two ships finally made wireless contact, the *Maryland* transmitted the same message 12 times, and in one repetition each word in the message was repeated three times. The *Colorado* was unable to decipher the message.

With the memory of that still rankling, Admiral Evans was in no mood for tolerance when his officers complained that the newly installed radio telephone was seriously hampering the operation of the not so good, but at least more familiar, regular wireless sets. The Admiral peremptorily ordered all radio telephone use discontinued for the balance of the voyage.

That word never reached the happy deForest until the ships had been six weeks out of Norfolk and had made the transit through the Straits of Magellan to Possession Bar in Chile. There was no radio telephone in use, but mysteriously a glowing account appeared in the newspapers which told how the almost miraculous use of the new technique had guided the Great White Fleet safely through the storm-tossed Straits.

That added another mystery, for the actual passage through the admittedly difficult Straits was made in almost perfect weather. There was nothing more serious than fog, which occurred on the last 30 miles of the 360-mile transit.

Many historians label the newspaper account as completely false. No one knows who placed it in the paper. But the sale of deForest stock was brisk.

The expensive fiasco with the radio telephones

aboard the Great White Fleet caused an abrupt halt in U.S. Navy interest in the new medium. Not for eight more years did the Navy return to the new oddity and investigate its possible benefits for the fleet.

Admiral Evans went ashore when the Great White Fleet reached San Francisco on May 8, 1908, and he retired from the Navy. The big battleships and their auxiliaries continued sailing westward, until they completed a round-the-world journey in February, 1909, steaming 45,000 miles in 432 days. But the radio telephones were still silent.

Nothing can take away from the glory of that voyage, but one wonders what might have been the impact on world communications if the Admiral had been less irritable and less nervous and had given the sets a fair test.

Lee deForest wasn't crushed by the abrupt dismissal of his radio telephone by the Navy after the Great White Fleet test in 1907 collapsed. The following year he sailed for France, most of his baggage consisting of a radio telephone transmitter that he kept carefully in his cabin. In Paris, he told the delighted officials of his plans for a new experiment, and the next day joined the throngs of tourists going up in the triple elevators to the top of the Eiffel Tower. He hooked up an antenna, the predecessor for scores of others carrying radio and television beams, and began voice broadcasts that were heard as far as Marseilles, 500 miles to the south.

When he tired of speaking into the clumsy microphone, he turned on a phonograph, as Fessenden had done before him in New England, and sent music rippling over the air waves of France.

Only a very few people heard the results, but those were important. Within a short time deForest, cocky and self-assured, was sitting down with Italian officials, boldly proposing to equip the Italian Navy with his radio telephone.

The Italians shared the dread of all seamen—of close maneuvers in fog, of the need for split-second communications between commanders in times of urgency. They agreed, hesitantly, to try out two sets of the deForest equipment. Disappointed and nearing the end of his money, deForest headed back to the United States.

It was after he returned from Europe that he began to use musicians regularly in his experiments conducted in midtown New York. In 1910 deForest coaxed the world-famed opera singer Enrico Caruso to sing into a radio microphone he had placed in the wings of the Metropolitan Opera House. Seeking an influential audience, which under ordinary circumstances might consist only of some Navy wireless operators in the School at the Brooklyn Navy Yard, deForest persuaded reporters from *The New York Times* and other papers to listen in through bulky microphones which they pressed tightly to their ears.

The reporters were uncertain how to file on the event. If it was historic, they were not aware of it. And their factual accounts were rather depressing. *The New York Times* said, "The homeless song waves were kept from finding themselves by constant interruption. The reporters could hear only a sharp ticking." Another paper chronicled for history, "The guests took turns fitting the receivers over their ears, and one or two of them thought they heard a tenor, but they were not sure."

The most devastating comment came from the reporter who wrote of the experience, "There was an operator somewhere carrying on a ribald conversation with some other operator, greatly to the detriment of science—and an evening's entertainment."

DeForest's fortunes began to skid again. One of his companies was in serious trouble. This time it was more than the familiar scramble for money, more than the familiar harassment of patent-infringement lawsuits.

In May, 1912, deForest and four officers of the deForest Radio Telephone Company were arrested by United States Marshals on charges of having used the United States mails to defraud investors in radio stock.

DeForest stood trial as a criminal. He was branded for a time as a swindler and a cheat.

Two of the directors in the deForest Company were found guilty of fraud after a six-week trial. DeForest and another director were absolved of guilt.

"Tireless and jaunty, never long cast down by endless setbacks" was an accolade given deForest. He thrived on the deluge of abuse. Somehow or other he managed, in his own eyes at least, to remain stalwart and unshaken. His enemies never let him forget that he organized over 25 different companies, most of them on shaky foundations, mishandled by people who were callous with irresponsibility. All of them invariably slid into bankruptcy.

He was a genius and a villain. He was accused of outright stealing of patents, not with the cloak-and-dagger approach, but with the crude schoolboyish method of making furtive sketches while visiting in a fellow inventor's laboratory.

He fought tooth and nail, and finally he emerged triumphant, his fame nailed to the title of a book he wrote about himself, *The Father of Radio*. At the time of its publication, only one of the early battlers was there to dispute his claim.

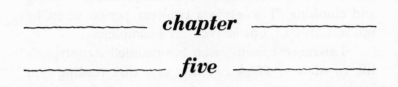

chapter five

One man, Edwin Howard Armstrong, was responsible more than any other for the superbly clear sounds that we hear from radio today.

When Armstrong first entered the scientific scene at Columbia University in 1912, there was no hint of the tragedy that was finally to overcome him.

He was a tall, gangling youth, prematurely bald. He was a genius in electrical research, and he embarrassed some of his professors with a blatant display of his overwhelming knowledge.

There was no scientific inheritance from Armstrong's forebears that would have foretold the striking career awaiting him. His father sold Bibles and other religious literature; his mother was a schoolteacher. Armstrong was born in 1890. His early years in staid New York City homes were remarkably calm. He was part of a large family that swelled a quiet city neighborhood.

Normalcy was the keynote for Armstrong during his childhood. He delighted in tennis, skating, running and climbing. There was a two-year period when he was seriously ill, but he recovered completely.

Lawrence Lessing, who has detailed Armstrong's life in *Man of High Fidelity,* says that reading two books on inventions gave Armstrong his firm desire to be an inventor. When he entered high school he was already committed to a career in wireless—a commitment which, in most youngsters, would be as fleeting as the wind. But Armstrong stayed with his ambition.

He picked as a guiding light Michael Faraday, who had discovered electrical induction, the principle of the dynamo upon which was laid much of the foundation for the electrical power industry, nearly three quarters of a century before. Another guiding light for Armstrong was the bold, dashing, colorful Marconi. He was fascinated by the inventor's fast automobiles and sleek yachts.

Some hero worshippers would be content with reading of the exploits of their heroes. Armstrong did more than that. He joined the ranks of amateur wireless operators.

The amateurs came to the surface at the turn of the century when Marconi performed the first of his striking wireless feats. In the beginning the enthusiasts who assembled their own wireless sets numbered only in the hundreds. It took a special kind of youngster, one with an aptitude for things scientific, to assemble the complex circuits. And the sets were expensive. An office boy making $5.50 a week had to think twice before he embarked on the new hobby. To build a set took far more money than raising pigeons.

The youngsters spoke a language of their own—not only the language of science, but a mysterious language of the Morse code's dots and dashes. They "spoke" from one attic studio to another several miles away; they "spoke" in *dit-daws* from one barn to another barn. But most of all, they listened eagerly to the chattering conversations that went on uninhibited all about them, a flood of electronic impulses that they could understand and interpret just as easily as the intended recipient. The open skies, filled with unseen signals, were one vast party line. Anyone with the necessary skill and equipment could listen in.

At first the messages went from one backyard to another, then from neighborhood to neighborhood, then from one city to a nearby city. As the amateurs became more skillful, the equipment improved. Youngsters sitting in their bedrooms with earphones clamped to their heads began to reach out, first to 500-mile contacts, then even to an occasional 1000-mile two-way contact.

They thrilled to calls for help from sinking ships; they followed, message by message, the progress of rescue attempts.

When that palled, the amateurs were not above creating synthetic excitement. A few of them began sending out false messages that caused international distress, confusion and waste of time and resources. Coast Guard and Navy vessels were sent scurrying about in search of nonexistent ships that were about to "sink." They were part of the general confusion— jamming the wave lengths, interfering with each other, with the ship-to-ship commercial traffic and with the Navy signals. Radio was born in chaos and interference. The air waves were a jungle.

But these events should not outweigh the contribution of the young people scattered all over the country, drawing dots and dashes and occasionally words and music from the air all about them. They were the strong cadre of interest that kept alive the excitement of wireless and, later, radio.

If the amateurs had done nothing but give Edwin Armstrong a stage on which to play out his role, that alone would have justified their carefree existence.

Armstrong, tucked away in an attic room of his father's Yonkers, New York, home started to collect spark coils and interrupters and coherers that were the mark of trade for the innovators. All during his high school days he punctuated his studies with endless hours at the key of his wireless sets. He gave himself wholeheartedly to his hobby.

He pursued it aggressively, building a wooden antenna, strong, wide and sturdy, adding one piece of lumber after the other until the contraption soared 125 feet in the air. When it was completed and hooked up to his collection of coils and coherers, he was able to hear, very faintly, the wireless signals broadcast from the Navy station at Key West, Florida.

By that time Armstrong was already enrolled in Columbia University, where he was alternately a thorn in the side and a particular joy to his engineering professors. According to Lessing, Armstrong's biographer, he embarrassed one professor by risking possible electrocution while demonstrating that the instructor was entirely wrong in his preachments on certain electrical phenomena. Others were intrigued by Armstrong's brilliance. They did much to calm his brashness and to mature his thinking.

There was nothing accidental about Armstrong's discovery in 1912, but it was to revolutionize the industry.

Armstrong's attic activities were focused on one aim—to improve wireless reception. Joined with that quest was another—to reach out over greater and greater distances. There was nothing unique about the quest. Practically every other amateur and professional alike was involved in the same search. The difference with Armstrong was that he succeeded. He solved a mystery that had been baffling the nation's leading scientists and inventors for more than a generation.

From the very beginning, wireless research signals had a sudden, mysterious way of ebbing and vanishing. Sunlight seemed to kill them almost entirely. The daylight signals were so weak that only a trained ear, pressed tightly to an earphone, could distinguish the whispering *dit-daw*s. The slightest interference wrecked reception. Commercial stations, depending on the dollar revenue from the dots and dashes, were constantly at war with the amateurs who rode ruthlessly into the same wave lengths, causing havoc with the commercial messages.

Armstrong, as part of his experiments, put the deForest Audion tube under a microscope and noticed what the older man had missed. By jiggering some wires and placing a second tuning coil in the circuit, he got an immediate increase in the amplification. He worked alone in great secrecy during the winter of 1912-13, with astonishing results. Signals from distant stations that had been just a whisper when received under the most favorable conditions came in so loud that the headphones could be placed on a table, and the sharp sound would fill the room.

His receiving set picked up nearby amateur stations with almost cannonlike volume. His tuning coil, combined with deForest's Audion tube, was enabling him to reach across the ocean and pick up clearly signals from stations in Europe. Turning west, he pulled in San Francisco and Honolulu.

The Audion tube, one of the most controversial inventions in American history, bears close scrutiny.

DeForest had devised it very early in his experiments, then had gone back and reviewed it seriously just after Fessenden's feat of broadcasting the human voice in 1906.

DeForest made changes, notably when he devised a small wire grid within the tube which he had fashioned for him by a maker of Christmas tree light bulbs. The grid acted on electric currents much like a throttle on a locomotive, setting powerful currents in motion or slowing them down. The signals were heard with great clarity. There was no lag, no distortion.

But whose tube was it?

In broad outlines, for it generated a complex patent situation that reverberated in courtrooms for years, the background of the controversial Audion tube is this: Thomas Edison, while experimenting with his incandescent lamp, the electric light bulb, stumbled on the basic principle of today's electronic vacuum tube. Twenty years later, in 1904, Ambrose Fleming, a physicist who had worked side by side with Edison but who now was with the Marconi Company as a consultant, reworked Edison's tube so that it could be used to detect wireless signals. This was the Fleming valve. It heard signals in the same strength in which they were received. It did not amplify them.

DeForest states in his autobiography that he came across the idea for the tube when he noticed during experiments in his attic bedroom that unseen electric waves emanating from his wireless sending set seemed to cause the old Welsbach gas mantle to ebb and flow in cycles of varying brilliance. Whatever his original reason, it is a matter of record that deForest added the tiny grid of fine wire to the existing Fleming valve in 1906, a simple, significant move that gave additional clarity and added strength to the signal. Signals too weak to be detected by the Fleming valve came across clearly with the deForest Audion tube. It could be used in multiples—10, 50, 100, even 300—to boost the strength. A series of deForest tubes could magnify the ticking of a watch until it sounded like a hammer blow.

The tube has been labeled "the little giant." It has been stated flatly of the Audion tube that it was "the one thing that made radio telephony a finished product."

But there are many who looked upon the Audion tube as another step in a score of steps. Dr. Irving Langmuir, General Electric Company physicist and chemist, and Dr. H. P. Arnold, a researcher in the same company, for example, both added significant improvements to the deForest tube.

That was the tube—but it was Armstrong who put it to work.

It was incredible, impossible, what the young Columbia University student did—but it was a fact. His one addition, the regenerative "feedback" circuit, had finally pulled the signals out in the open. Europe was only an electronic breath away.

All this, of course, referred to Morse code signals. Two more years were to pass before transmission of the human voice across the Atlantic.

Armstrong continued his experiments, alternating study with laboratory work at Columbia. He tried sending signals with the same regenerative system, using the new tuning coil and the deForest tube. With that simple movement, he had a transmitter that made sustained and reliable radio voice broadcasting possible. Lawrence Lessing says that Armstrong's new instrument marked the end of the dark ages in wireless telegraphy.

The 22-year-old Armstrong did not keep detailed notes, and he lacked Lee deForest's grim determination to leave no patent stone unturned. That carelessness and failure to record data eventually was to cost him dearly.

Armstrong was not the only scientist whose life was affected adversely by the unhealthy climate that surrounded those working with wireless. There was little communication between individual investigators, groups or companies.

Only a few generations before, scientists had been magnanimous, dispersing their ideas freely to their cohorts. Now the atmosphere was changed completely. A scientist worked in the laboratory with a patent application at his elbow. A scribbled date or a notorized statement establishing prior patent rights became more important than the invention itself.

Fortunes were waiting for the right claimant. The pursuit of that fortune by men, each one of whom was sure in his heart that he was rightfully entitled to it, was to make villains out of heroes and heroes out of villains.

At the time of Armstrong's discovery he was living on $600 a year as a teaching assistant at Columbia. The leaders of the communications industry asked for demonstrations of his new regenerative circuit. The Marconi people came, along with American Telephone and Telegraph Company officials and Lee deForest.

Armstrong was friendly with his visitors. He gave no technical details, but he did allow them to view his apparatus.

What deForest saw was enough to send him back to his laboratory, leafing through sheaf after sheaf of dust-covered notes. Dimly he remembered filing a patent on a process where he had hooked together several of the Audion tubes. DeForest said nothing, but immediately filed additional patent claims that interlaced tentacles all through the area covered by young Armstrong's efforts.

Then deForest entered suit for patent infrigement.

The suit was placed amid hundreds of others and attracted little attention.

The inventions came thick and fast. Who owned what? It is impossible to simplify the situation, for the inventions were so entangled that lawyers for generations quarreled over the infringements.

The public was jaded and skeptical. Already the young wireless industry was tagged as one rich with shysters and charlatans, paced by fast-talking promoters anxious to lure the unwary public into buying worthless stock.

DeForest's suit for patent infringement preyed on Armstrong, although it did not deter him from continuing his scientific experiments. He invented other highly important circuits that were to lift radio to superb perfection. More than any other man he was responsible for bringing high-fidelity sound to the listening world. But like Fessenden, he never gained the fame accorded Marconi and deForest.

He was better known to lawyers, for half of Armstrong's life was spent in court in some of the longest, most notable, and most acrimonious patent suits of the year. He made millions of dollars from his patents, and he spent most of it fighting for his patent rights.

He hated Lee deForest, with whom he was engaged in a marathon legal battle that lasted 20 years and was not settled until the United States Supreme Court made the final decision. DeForest won. Armstrong stepped from a window in a high building and plunged to his death.

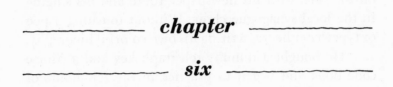

chapter

six

Marconi, deForest, Fessenden and Armstrong, the inventors, had prepared the equipment. Now another type of genius was required to turn the weird collections of tubes and wires from laboratory curiosities into compact devices that would go into the homes of millions of people throughout the world.

David Sarnoff was that kind of a genius.

He rose from an office boy to a corporate genius who had a direct influence on shaping world-wide communications.

He came to the United States in 1900, and he shined shoes, ran errands and sold newspapers. He stumbled into his lifework by accident. He meant to apply for a job as a copyboy on a New York newspaper, and he inadvertently knocked on the wrong door—that of the Commercial Cable Company, which was linked by undersea cable to all parts of the world.

Instead of rushing galley proofs, he started pedaling a bicycle, delivering cable messages. He continued, on the side, with his newspaper route and his singing in the local synagogue choir. Without touching a pen or typewriter he was writing his own colorful biography.

He bought a dummy telegraph key and a Morse code book and began to practice in the few hours of free time that were available in his busy life. He bought books on telegraphy and electricity. He cultivated friendships with the regular operators and persuaded them to let him talk to distant operators.

Sarnoff turned tragedy into triumph. He was fired from his Cable Company job because he insisted on giving priority to a synagogue choir assignment. He walked the streets of New York, then popped into the small, cluttered offices of the American Marconi Company, the insignificant beachhead that had been quietly operating for seven years at 27 William Street.

On that momentous day, in September, 1906, he was just past 15 years old. His entrance through the small door should have been heralded by the roll of drums, for that was the beginning of a lifelong career for both Sarnoff and the wireless company. Sarnoff was never to lose his driving force or his identity; the wireless company, by the strange, mysterious workings of finance, would lose its identity while slipping over into the American scene, but it still lives, buried deep within the corporate body of today's Radio Corporation of America.

The big step upward for a telegraph operator was to serve aboard one of the ships using the Marconi system. Unfortunately, in 1907, at the time Sarnoff came over the horizon, there were only four ships using the system. But a wireless operator was taken ill, and Sarnoff, resplendent in a new uniform, holding himself erect, went to sea as a 16-year-old telegraph operator. The round trip to Europe lasted only three weeks, and Sarnoff was again tied to a set in the New York City office. But the world had opened up for him.

He went next to the bleak, lonely wireless outpost on Nantucket Island, Massachusetts, where he turned the loneliness into a one-man 18-month university session in which he crammed into his thirsty brain every bit of technical information about wireless communication that he could assimilate.

He was only 18 years old when he was appointed manager of a Marconi station near New York City; a year later, he shipped with a seal-hunting expedition north toward the Greenland Sea. It was the first time a wireless operator had ever gone along on the lonely, perilous hunts.

It was a great maritime disaster that brought Sarnoff into the spotlight. Like most of his applauded appearances in the half-century to follow, the attention he received was merited.

At 2:20 A.M. on April 15, 1912, the British White Star liner *Titanic,* at that time the largest ship afloat, went to the bottom of the Atlantic after striking an iceberg at full speed on her maiden voyage from England to America. Of the 2224 persons aboard, 1513 were lost.

At that time, Sarnoff was working as the manager of the Marconi wireless station situated in the Wanamaker Department Store in New York City. He was on duty when, very dimly through the customary flood of dots and dashes that filled the air, he picked out a cryptic message from the steamship *Olympic*, 1400 miles at sea. "S.S. *Titanic* ran into iceberg. Sinking fast."

Sarnoff reacted quickly. He acknowledged receipt of the message, asked for more details and notified the press—the first the world knew of the unfolding disaster. Some of the most notable figures in the world were aboard the big ship, the most luxurious of its time. Crowds began to converge on the department store and on the 21-year-old Sarnoff, who sat by his wireless key, only the tireless movement of his skilled fingers betraying the unseen drama that was being enacted. The crowds were held back, and he kept "talking," maintaining the thin line of communication with the ships that were speeding to the lifeboats containing 700 survivors of the disaster.

Other wireless stations came charging on the air, reaching out with electronic fingers to query and demand and advise and threaten. The interference became so grave that contact was almost lost with the rescue ships. In a quick, decisive movement, President William Taft ordered all other wireless stations to close down so that Sarnoff could continue unimpeded his contact with the rescue operations.

He was at the key for 72 hours without sleep, while the entire world waited for the list of names that were spelled out in dots and dashes. When the name of the last survivor had come across, Sarnoff stepped back from the set.

Those three days stamped the worth of wireless on the minds of the entire world. They catapulted Sarnoff higher along the road that he was to follow in corporate pre-eminence in American communications.

As the British-owned Marconi Company settled more firmly on the American scene, Sarnoff left the wireless key and started the climb upward in the company. He was still a youth, yet he was exerting a major influence on the wireless company.

In 1916 he wrote a letter to a vice-president in charge of operations:

> I have in mind a plan of development which would make radio a "household utility" in the same sense as the piano or phonograph. The idea is to bring music into the home by wireless.
>
> While this has been tried in the past by wires, it has

been a failure because wires do not lend themselves to this scheme. With radio, however, it would be entirely feasible.

For example, a radio telephone transmitter having a range of say 25 to 50 miles can be installed at a fixed point where the instrumental or vocal music or both are produced. The problem of transmitting music has already been solved in principle and therefore all the receivers attuned to the transmitting wave length should be capable of receiving such music. The receiver can be designed in the form of a simple "Radio Music Box" and arranged for several different wave lengths, which should be changeable with the throwing of a single switch or pressing a single button.

The "Radio Music Box" can be supplied with amplifying tubes and a loudspeaking telephone, all of which can be neatly mounted in one box. The box can be placed on a table in the parlor or living room, the switch set accordingly and the music received. There should be no difficulty in receiving music perfectly when transmitted within a radius of 25 to 50 miles.

Within such a radius there reside hundreds of thousands of families; and as all can simultaneously receive from a single transmitter, there should be no question of obtaining sufficiently loud signals to make the performance enjoyable.

It is not possible to estimate the total amount of business obtainable with this plan until it has been developed and actually tried out; but there are about 15,000,000 families in the United States alone, and if only one million or seven percent of the total families thought well of the idea it would, at the figure mentioned, mean a gross business of about $75,000,000 which should yield considerable revenue.

Aside from the profit to be derived from this proposition the possibilities for advertising for the company are tremendous, for its name would ultimately receive national and universal attention.

The idea, in 1916, was incredible. Voice radio communication was still hopelessly bogged down in the concept of person-to-person messages. Music for the millions was preposterous. Sarnoff's carefully constructed letter was tossed aside and completely forgotten for five more years.

The British directors, working behind the screen of the "American" labeling of the Marconi Company, were in a strong position and getting stronger with time. One of the deForest companies, which had equipped more than 200 ships with its system, had slid into bankruptcy, and the British company leaped to meet the needs for equipment. Other American companies raced on the scene, collided with financial reality and folded abruptly. In 1916 there were 582 ships at sea equipped with wireless apparatus licensed by the British.

The prospect of a supercompany, controlling a world-wide system, became a reality.

The parent Marconi Company denied all evil intent, but the danger was there nevertheless. And the danger became ever more apparent when the company started building a transatlantic wireless station near Belmar, New Jersey.

The American Telephone and Telegraph Company was not about to let the invasion go unchallenged. The transcontinental telephone link between New York and San Francisco had finally been completed in 1915, using the deForest Audion tube. Now AT&T was turning wholeheartedly to radio telephony, and the Audion, coupled with the Armstrong feedback system, was again to be the key. The takeoff spot for the AT&T experiment was the Naval station at Arlington, Virginia, where three towers, one of them 600 feet high, the other two 450 feet high, stabbed into the air.

Early in the morning of September 29, 1916, Theodore Vail, president of AT&T, spoke into his long-handled desk telephone. Land wires carried his words to the towers of the wireless station at Arlington, Virginia, where they were delivered to the sending apparatus of the wireless telephone. Leaping into space, they traveled in every direction. The antenna of the wireless station at Mare Island, California, in San Francisco Bay, caught some of the waves, and one of the AT&T engineers heard the voice of his chief and spoke to him at length.

The waves sped on and were intercepted by another AT&T engineer stationed at Pearl Harbor near

Honolulu, Hawaii. The distance was nearly 5000 miles.

Three weeks later, on October 21, 1916, the words were beamed eastward over the Atlantic Ocean and were heard with great clarity by the station located on the Eiffel Tower in Paris. Fessenden had spoken over space nine years before, but now the Atlantic was bridged, and the words rode on a strong, unswerving beam that carried the human voice without faltering. More than 300 vacuum tubes, ranged in great banks of power, gave the impetus to the voice signals.

It was a gripping demonstration. The deForest Audion tube and the disputed Armstrong regenerative circuit changed the whole aspect of radio telephony from a hit-or miss curiosity into a reliable system of radio communications.

The new day had dawned.

But what kind of a day was it?

Even the giants like Marconi and AT&T were confused about this new prodigy.

It was gauged always as a competitor of the established dot-and-dash wireless signals.

One official of the Bell Telephone Company, nervously assessing the possible competition of "radio" (though it was not yet known by that name), said, "It is difficult to see where the [radio] business can ever be successful or profitable." The biggest handicap, the telephone company believed, was the utter lack of privacy in transmitting messages by the radio method. Anyone with a receiving set could listen.

Walter Kellogg Towers, who surveyed the wireless telephone (radio) scene in 1917 in his book, *Masters of Space,* said very firmly:

> It is natural that one should wonder whether the wireless telephone is destined to displace our present apparatus. This does not seem at all probable. In the first place, wireless telephony is now, and probably always will be, very expensive. Wherever the wire will do, it is the more economical. There are many limitations to the use of the ether for talking purposes, and it cannot be drawn upon too strongly by the man of science. It will accomplish miracles, but must not be overtaxed. Millions of messages going in all directions, crossing and recrossing one another, as is done every day by wire, are probably an impossibility by radio telephony. Weird and little-understood conditions of the ether, static electricity, radio disturbances, make wireless work uncertain, and such a thing as twenty-four-hour service, seven days in the week, can probably never be guaranteed. In radio communication all must use a common medium, and as its use increases, so also do the difficulties. The privacy of the wire is also lacking with the radio telephone.

The nation had the radio telephone as an accomplished fact. What would it do with it?

The decision was taken out of the corporate board rooms when the nation edged closer to war in 1917.

The possibilities of wireless, in those days when a German spy lurked around every corner, were electrifying to the imagination. As a precaution against any stray signals creeping over to Germany, all amateur wireless stations were ordered dismantled. With wartime zeal the military police, the private police and the private citizens responded enthusiastically. There was much smashing and locking, and tearing down. Even the massive steel tower Reginald Fessenden had built at Brant Rock, Massachusetts, the $1,000,000 memorial to his determination to shove the human voice into space, was torn down lest it be used for surreptitious messages to U-boats lurking far out at sea. The tower, in a crashing dive, became 60 tons of iron and steel. It was sold for junk.

Early in the war, when the United States had proclaimed its neutrality, any exchange of wireless messages between belligerent nations and stations in the United States was forbidden. The American Marconi Company ignored the order and handled messages from the British fleet cruising off New Jersey. The American government reacted swiftly, shutting down the station for four months until the Marconi Company agreed to the censorship. The Marconi Company's high-handed approach in dealing with the Government would be remembered.

World War I was of inestimable value to the future of radio broadcasting, not because of any startling demonstration of new equipment but because the Government, in one decree, swept all the patent litigation under the rug, called in the warring companies and abruptly declared peace. Westinghouse, General Electric, Marconi, the United Fruit Company —which had its own nest of patents—American Telephone and Telegraph and Lee deForest suddenly found themselves side by side, not in a courtroom but in common ventures leading to Allied victory. The very best possible wireless sets were built for military use, a slick cannibalizing of the best of ideas and patents from each of the companies.

The idea worked to perfection. But then the war ended and the legal nightmares started again. The baby industry was prepared to commit a kind of suicide. The Marconi Company stepped in and, with a series of swift maneuvers, was soon in a position to straighten out the entire mess.

It was an innocent enough proposal; Marconi wished to get exclusive rights on a General Electric alternator, Dr. Alexanderson's unit that had originally launched Reginald Fessenden's voice into space.

Great Britain already controlled the underwater cables that carried telegraph messages between the continents. Marconi had to be stopped. Otherwise America would always be looking over the fence, dependent on Great Britain for life lines of communication.

The thought of radio communications resting solely in the hands of "foreigners" was enough to cause apoplexy in the United States military departments. A move was started to nationalize the entire radio communications industry, with the Government in full control.

By this time the bulk of the important radio patents had slipped into the hands of the American Telephone and Telegraph Company, General Electric, Westinghouse and the Marconi Corporation. Prodded by the United States government, Marconi suddenly found itself out in the cold, with the three American giants joined in uneasy temporary partnership. The solution was the elimination of the British from the American communications scene, an elimination accomplished with the abruptness of a guillotine.

The move was all very legal. General Electric paid cash for the British-owned shares in the American Marconi Company.

What of Sarnoff? He and all the top-echelon managers came along in a package deal. The only outward change, for them, was the sign on the building denoting the new name, the Radio Corporation of America.

That was only the beginning for RCA. Into the

new company came Westinghouse, bringing with it all the companies it had previously bought—including the National Signalling Company, complete with the Fessenden patents. The United Fruit Company and its patents also merged into the operation.

All the patents were tossed into a common pot, like the ingredients in a stew. The upshot was the Radio Corporation of America, formed first as a kind of super shoe store where the products were the bits and pieces that could go into the making of a radio receiving or transmitting set.

RCA had another important contribution. It brought to the fore Dave Sarnoff, blessed with a genius for exploiting a new industry.

By October, 1921, the maneuverings were completed. Westinghouse and General Electric cut up the manufacturing pie—60 percent of the radio equipment to be built by General Electric, 40 percent by Westinghouse. The new RCA was to be the exclusive U.S. sales agent.

On paper, RCA, the "front" for General Electric and Westinghouse, had the entire radio communications business in the United States secure in a tight, patent-protected package.

It didn't work out that way.

chapter

seven

Radio historians agree that the irrascible Reginald Fessenden was the first to send the human voice out over the airwaves.

Not so much unanimity exists as to the first radio broadcast. The deForest rooters point to Enrico Caruso's debut from the Metropolitan Opera House in New York as early as 1910. Then, to strengthen the claim of the prolific inventor, his supporters recount what appears to be irrefutable evidence: on November 7, 1916, deForest announced over the air the results of the hotly contested Presidential election when Woodrow Wilson was running for re-election against Charles Evans Hughes. Standing tall and unflinching before the microphone, not quite certain that anyone was listening, deForest read flash election bulletins at regular intervals throughout the exciting night.

The event was reported in *The New York Times* for Wednesday, November 8: "The Bronx produced an election night innovation when, shortly after dark last evening, the deForest radio laboratories in High-bridge began flashing the returns. Amateur operators within a radius of 200 miles had been forewarned of the new information service, and it was estimated several thousand of them received the news, many of them using the newly manufactured wireless telephones."

The event lost some of its luster, however, when deForest was swept up in one of the most painful mistakes in the history of the news media. He informed his unseen audience, "Charles Evans Hughes has been elected President of the United States."

Hughes, of course, was not elected. The Wilson victory emerged the next morning while deForest was still sleeping off his previous night's "triumph."

The coming of the war and the complete shutdown of all stations except the few authorized by the government brought a halt to the deForest "broadcasts."

The shutdown, the forced marriage of the various companies manufacturing wireless sets and the war itself were big factors in the fast-approaching radio explosion. Good equipment, embodying the best that was available, now was in daily use.

With a few rare exceptions, the war effort in radio telephone work was still directed toward person-to-person communications, more or less a stepsister of the accepted dot-dash wireless. The Germans, in the last

months of the war, did go on the air with voice broadcasts from their most powerful station in a Berlin
suburb. They read bulletins in German, French and
English telling of German successes, which were slim,
and minimizing German troop losses, which were considerable. They were the first of endless propaganda
broadcasts that were scheduled for the coming decades,
but the audience was limited to a handful of radio
operators.

Every step that was taken was a preparation for
events that were shortly to come.

With the end of the war and the lifting of restric-
tions on wireless stations, the claimants to "firsts" in
broadcasting began to appear. The United States Army
Signal Corps broadcast services from Trinity Church,
Washington, D.C., on August 24, 1919.

Shortly afterwards came the powerful United
States Navy station, NSF, which began broadcasting
music, with an occasional health lecture, early in 1920.

A commercial broadcast is one that is advertised
before it happens, and one that continues. For example: "Tomorrow night, between 8:30 and 9:30 P.M.,
we will broadcast dinner music. We will broadcast
again at the same hour with the same program each
succeeding night of the week."

Using that as a guideline, neither deForest, the
Signal Corps, nor the U.S. Navy can be cited for the

"first" in commercial radio broadcasting. That honor must go to Dr. Frank Conrad, and the pioneer Westinghouse station he fathered, KDKA in Pittsburgh, Pennsylvania.

Years of preparation went into earning the accolade for Conrad and KDKA. The lucky break came in the face of apparent disaster.

Westinghouse entered the radio telephone business through a special back door opened by both the British and the American governments. Conrad and the other Westinghouse engineers started to investigate secret aspects of radio telephone communications in August, 1916, at the request of the British Government.

Conrad was not a lonely Armstrong, a secretive deForest, or an angry Fessenden. He was a cool, hardheaded, hardworking scientist who had a job to do. He did not shut himself away from the world, but he worked in very close cooperation with the United States Signal Corps. Ample funds were supplied for the experiments. Conrad luxuriated in a big staff and a lot of good equipment.

The Westinghouse group was deep into research when the drastic order came along closing down every wireless station in the nation, amateur and professional alike, on April 8, 1917. For a brief period it appeared as though the Westinghouse development efforts on highly improved radio telephone and wireless sets for use on battleships would be swept away.

Fortunately common sense intervened, and the Westinghouse engineers were granted a special license to continue their experiments.

But overzealous neighbors, hearing the wireless signal, set the police on Dr. Conrad's trail as a possible spy. When the police swooped down, Dr. Conrad had slipped from their clutches. He was "discovered" sitting in the office of the Secretary of the Navy at Washington discussing the progress of his experiments for radio sets aboard battleships. The spy scare ended abruptly.

One of the Westinghouse stations was located at the plant at East Pittsburgh. The other, four or five miles away, was on the second floor of a garage at the Conrad home.

The ban on amateur stations was lifted on October 1, 1919. Almost overnight the military forgot that the two experimental stations existed, and Westinghouse was left with a collection of radio tubes and electronic circuits, all of them gathering dust. Urged on by Conrad, the company decided to pursue the radio investigation. Westinghouse had bought, for a large sum of money, the International Radio Telegraphy Company, which owned many of the law-entangled patents. This put Westinghouse on a collision course with General Electric, which had begun its maneuvers to buy out American Marconi. Fortunately the collision was avoided when the two big electric companies joined hands to form RCA.

But long before the marriage and the birth of RCA, Dr. Conrad was back in the garage, improving his radio telephone transmitter. Conrad's daily efforts were on a bona fide scientific level. He or his assistants sent out signals, chiefly discussions of the kind of equipment being used and the results obtained. They requested any amateurs who might happen to hear a discussion to mail in a report on the quality and strength of the reception. The scientists could then plot the efficiency of the transmitter and plan improvements.

It was a monotonous routine, and there was a danger that even the avid radio amateurs would get tired of hearing the serial numbers on a vacuum tube or a condenser and not bother to acknowledge receipt of the signal. To ease the burden on his assistants and to "hook" the interest of possible cooperators in the scientific experiment, Conrad, like Fessenden and de-Forest before him, placed his microphone before a phonograph and substituted music for voice.

The music saved Conrad's voice, and it delighted and amazed "hams" all over the country. Mail, heavy previously, became a deluge of requests that records be played at special times so that the writer might convince some skeptic that music really could be transmitted through space.

Specific requests were played as long as this could be arranged, but so heavy was the demand that Conrad was forced to announce that instead of complying with each individual request, he would "broadcast" records for two hours each Wednesday and Saturday evening. This is the first recorded use of the word "broadcast" to describe a radio service.

These broadcasts soon exhausted the supply of records. The Hamilton Music store in Wilkinsburg offered a continuing supply of records if Conrad would announce that the records could be purchased at the store. Conrad agreed and thus gave the world its first radio advertiser—who promptly found that records played on the air sold better than others.

This two-a-week program schedule was continued, with live vocal and instrumental talent added from time to time.

By late summer of 1920, interest in these broadcasts had become so general that the Joseph Horne Company, a Pittsburgh department store, ran this ad in the SUN on Wednesday evening, September 29, 1920:

AIR CONCERT "PICKED UP" BY RADIO HERE
Victrola music, played into the air over a wireless tele-

phone, was "picked up" by listeners on the wireless receiving station which was recently installed here for patrons interested in wireless experiments. The concert was heard Thursday night about 10 o'clock and continued about 20 minutes. Two orchestra numbers, a soprano solo—which rang particularly high and clear through the air—and a juvenile "talking piece" constituted the program.

The music was from a victrola pulled close to the transmitter of a wireless telephone in the home of Frank Conrad, Penn and Peebles Avenues, Wilkinsburg. Dr. Conrad is a wireless enthusiast and "puts on" the wireless concerts periodically for the entertainment of the many people in this district who have wireless sets.

Amateur wireless sets, made by the maker of the set which is in operation in our store, are on sale here $10.00 up.

Conrad's boss at Westinghouse was Vice-President H. P. Davis. The big assist that Davis now gave to Conrad adds fuel to the discussion: who contributes most to the progress of invention—the lone scientist, like Fessenden or deForest, who plugs away almost unnoticed and practically always unaided in a small laboratory; or the industrial researcher, perhaps with less genius than that of a Fessenden, but strengthened by the money, equipment and encouragement of a big corporation?

Davis is credited with the first vision of a new radio industry whose strength would be in the manufacture of home receivers and the supplying of radio programs which would make people want to own such receivers. The important thing is that Davis did display a commendable alertness. He exercised great influence in hastening the coming of the corporate profit bonanza, the deluge of American broadcasting. Here is how he describes the events leading up to Westinghouse's corporate plunge:

> We at the office were watching Dr. Conrad's activities at the transmitting station very closely. The advertisement of Horne's Department Store, calling attention to a stock of radio receivers which could be used to receive the programs sent out by Dr. Conrad, caused the thought to come to me that telephony as a confidential means of communication was wrong, and that instead its field was really one of wide publicity, in fact, the only means of instantaneous collective communication ever devised. Right in our grasp, therefore, we had that service which we had been thinking about and endeavoring to formulate.
>
> Here was an idea of limitless opportunity if it could be "put across." A little study of this thought developed great possibilities. It was felt that here was something that would make a new public service of a kind certain to create epochal changes in the then accepted everyday affairs, quite as vital as had been the introduction of the telephone and telegraph, or the application of the electricity to lighting and to power. We became convinced that we had in our hands in this idea the instrument that would prove to be the greatest and most direct means of mass communication and mass education that had ever appeared. The natural

fascination of its mystery, coupled with its ability to anni-
hilate distance, would attract, interest and open many
avenues to bring happiness into human lives. It was obvi-
ously a form of service of universal application, that could
be rendered without favor and without price to millions
eager for its benefits.

Davis persuaded Westinghouse management to
earmark company funds for a broadcasting station to
be built atop the Westinghouse factory in East Pittsburgh.

It took time to gather the necessary equipment,
and it was late in the fall of 1920 before the station
was ready for operation. There was a reason for the
slowness. Every step was new. No one had ever before
built a radio station with the avowed purpose of enter-
taining an unseen audience.

Westinghouse enlisted the aid of the local newspapers, and a series of stories began to appear, promising radio broadcasts from the Westinghouse station in
the near future. Few of the Pittsburgh readers could
comprehend the meaning of "radio broadcasts." Despite that fact, the initial newspaper stories served to
whet the appetites. Something was about to happen.

What was needed, Davis determined, was something spectacular—some special event that would
bring broadcasting on the air in a tumultuous burst of
publicity.

He found the spectacular: the 1920 Presidential

election, the unequal struggle between Warren Harding and James Cox, would be the opener. Station KDKA would open on November 2, 1920, broadcasting the election results.

A radio station is a useless enterprise unless there is someone to listen to it.

Who would be listening when KDKA went on the air? The question had tormented Fessenden 14 years earlier. It still torments broadcasters today.

The Westinghouse people were somewhat in the position of a man who builds a big ship not knowing if there will be water in which to float it. They might plan to send out the election results on schedule, but who would listen? Was anybody out there? There were amateurs with small sets, but they were a rather unstable, erratic audience. Most of them were more interested in dot-dash conversations with fellow amateurs than in any tinny musical notes that might come straying into the frequencies, rendering dots and dashes unintelligible.

It was estimated that there were 30,000 amateur wireless operators in the country at the beginning of 1920.

Westinghouse Vice-President Davis, however, didn't know the number at the time, or if he did, he didn't know how many of the amateurs were available in the Pittsburgh area.

He decided to nail down a captive audience for

the planned pioneer broadcast of the Presidential returns.

Davis requested his engineers to manufacture a number of simple receiving outfits and distributed them among friends and officers of the company. It was a needless chore. When news of the company's plan to broadcast the election results was released in the Pittsburgh newspapers, the entire city, so it seemed, was caught up in a radio fever. The shops were denuded of their stocks of electrical accessories for do-it-yourself receivers. Anxious amateurs traveled to other towns to search for parts sold out in Pittsburgh.

In a wooden shack atop the tallest factory in the Westinghouse East Pittsburgh complex, on November 2, 1920, a switch was turned and broadcasting began.

A single room accommodated transmitting equipment, turntable for records and the first broadcast staff: William Thomas, operator, and L. H. Rosenberg, announcer. R. S. McClelland and John Frazier handled telephone lines to the newspaper office.

Gleason L. Archer, in his authoritative *History of Radio,* says of the momentous pioneer broadcast,

> [The wooden shack had] two windows. The transmitter was in the corner between the windows and was flanked by table-like desks, set against the wall on two sides. The engineer sat before the transmitter and the announcer was

stationed close by. The microphone seems to have been much like an old-fashioned telephone set with a box-like arrangement behind the mouthpiece. The hastily constructed station had two 50-watt oscillators and four 50-watt modulators.

Headphones were worn by the engineer and the reporters, two of whom were present during the broadcast. So hurriedly had the work been rushed to completion that there had been little opportunity to test the sending apparatus before the evening of election day. On the night of November 1st it was tried out but the results were unsatisfactory.

Adjustments were made next day. Conrad was very nervous over the prospect, so nervous in fact that he rushed home to his own garage-station and remained there during the broadcast, "standing by" to pick up the program in case anything should go wrong at the new station.

Broadcasting began at 6 o'clock election night and continued until noon the following day, even though Candidate Cox had conceded the election to Senator Harding hours earlier.

All through the stormy night the signals went out from the infant station. As though in ignorance of the new miracle, or contemptuous, the usual crowds stood in the driving rain before outdoor billboards in downtown Pittsburgh.

Over and over again in between election returns and occasional music, the station announcers asked plaintively, "Will anyone hearing this broadcast communicate with us, as we are anxious to know how far the broadcast is reaching and how it is being received?"

To the astonishment of Westinghouse, thousands of people in their homes and in the shops of enterprising electrical dealers, who had set up receivers for the occasion, had tuned into the broadcast. A flood of letters poured into the station, a response that prompted Westinghouse to announce that regular broadcasts would continue every evening between 8:30 and 9:30.

The big boom was still 18 months over the horizon, but radio was on its way.

Years later the radio industry dedicated a plaque at Dr. Frank Conrad's former home in Wilkinsburg in the suburban Pittsburgh area.

It reads:

BIRTHPLACE OF RADIO BROADCASTING
Here radio broadcasting was born.
At this location, Dr. Frank Conrad,
Westinghouse Engineer and Scientist
Conducted experimental broadcasts
Which led to the establishment of
KDKA and modern radio broadcasting,
And to the world's first scheduled
Broadcast, November 2, 1920.
                    Dr. Frank Conrad
                    1874-1941

chapter

eight

For the first heroic year of radio broadcasting there was a return almost to the pioneering days of lifesaving cunning and ingenuity in America. For a brief time money was not the key to radio listening. It was determination, resourcefulness and patience to absorb an almost meaningless gobbledygook that opened the doors to the new fascination that was gripping the land.

The stocks of "wireless" sets offered by Horne and Company in Pittsburgh at $10 were extremely limited, and even those simple sets required a bit of tinkering in order to get them assembled. Worse yet was the effort needed to assemble a set from scratch. The first parcels of components were incomprehensible to the average person. Advertisements specialized in bargain offerings of Kenotron rectifiers, oscillation transformers, transmitting grid leaks, filter reactors, magnetic modulators and triode amplifiers. The man in the street, in desperation, turned to the amateur and asked him to build a set. Swamped with requests, many of the amateurs started small manufacturing operations—a vague hint of the policing tasks facing the Radio Corporation of America, which thought itself to be the sole sales agent for many of the indispensable radio parts.

After the first mystery had worn off the electronic circuits and after radio antennas had begun to sprout like trees from rooftops, more and more people who hadn't the slightest scientific bent began the intriguing task of winding thin copper wires around an empty oatmeal cereal carton, the first step in constructing a homemade receiving apparatus.

It was one of the most exciting periods in American communication annals. A fundamental instinct had been touched. The thin, sometimes unintelligible voices and music coming through earphones were almost hypnotic in their mass effect on the American people.

The daily KDKA broadcasts created tremendous interest, not only in the hilly Pittsburgh area but nationwide. The station distributed mimeographed sheets of papers telling of the events projected for the one-hour broadcast in the coming evening. Soon newspapers all over the United States and throughout Canada were asking for copies of the schedule. Every night more and more Americans hunched over crude sets, hot earphones pressed tightly to their ears while they turned and announced in awed tones to those clustered about, "It's Pittsburgh!"

From the very beginning there was something electrifying about the response of people to this new "universal speaking service." H. P. Davis sensed the unseen turbulence barely three months after the first broadcast. His mind leaped beyond the crude realities of broadcasting and reception when he wrote, "It is not unreasonable to predict that the time will come when almost every home will include in its furnishings some sort of loud-speaking radio receiving instrument, which can be put into operation at will, permitting the householder to be in more or less constant touch with the outside world through these broadcasting agencies —the development will mark one of the great steps in the progress and evolution of mankind."

Can anyone imagine a similar accolade being made about the future of the airplane three months after the Wright brothers lifted from Kitty Hawk? Would the same ringing prophecy have been made about automobiles in December, 1893, three months after the Duryea brothers had rolled their pioneer automobile out on the dirt roads of Massachusetts?

But Davis was more than a visionary. He was anxious to get radio receiving sets into the living rooms of the nation.

Thanks to Davis, Westinghouse was ready with a

receiver shortly after KDKA went on the air. The price was approximately $125, a price that was scorned by Americans who were busy building sets around oatmeal cereal cartons. The mass market reaction to the initial offering was a dull thud.

Westinghouse moved quickly to undercut the oatmeal carton competition. The company engineers concentrated on a receiving set simple enough for any nontechnical person to operate and, most important, cheap enough to elbow the oatmeal builders out of the way.

In June, 1921, the company brought on the market the Aeriola, Jr., a tiny crystal set in a steel box six by six by seven inches in size. It had a range of 12 to 15 miles and sold for $25.50, complete with headphones, antenna equipment and "full instructions." For the affluent there were more elaborate crystal sets for $32.50 and $47.50. The Aeriola, Jr., was a smash success.

What chased the humble crystal set out of the living room was an announcement by RCA of a revolutionary new vacuum tube, one that used far less current than previous tubes and had an extremely long life. The unreliable tubes of pre–World War I days had cost $50 and lasted only 70 hours. These new tubes, incomparably better, cost only $7, and the price was reduced rapidly to $1.50. With each price cut the life of the tube, through improved engineering, was extended. Eventually it reached 5000 hours.

On the heels of the tube announcement, Westinghouse brought out a one-tube set that sold for $75, complete with batteries and antenna. Snob appeal worked in sending the cumbersome, dust-collecting oatmeal carton crystal sets into obscurity. This was the beginning of a period when one's social status was almost determined by the number of tubes in one's table radio. General Electric came out with a three-tube set, and Westinghouse bounced back with a four-tube set. Prices up to $400 were quoted—the equivalent of four months' wages for most people at that time.

Loudspeakers also came into general use, relegating the uncomfortable earphones to the museum shelf. In the beginning the only loudspeaker available was similar to a brass automobile horn with a telephone receiver on the end; it sold for $40. Phonograph attachments, permitting use of the phonograph as a loudspeaker, sold for $18. Both of these "loudspeakers" were headed for early extinction.

The number of dials and controls on the sets was awesome. In the earlier, crude models, three critical adjustments were needed—tuning, regeneration control and filament control. The expensive sets managed to get an additional two dials on the face of the set so that it took on the aspects of a control panel in a jet liner cockpit of today. Turning on the radio set was not a casual operation. There was generally a key member of the family who was called upon to exercise dialing judgment.

When RCA brought Edwin Armstrong's super-heterodyne circuit to the market, much of the need for the "cockpit" controls vanished because the signals were then so strong that it required no special skill to set the dials correctly. The super-heterodyne also did away with the need for an outside aerial, but so strongly was the public wedded to the wires stretching tautly from point to point on the rooftop that many kept the unneeded aerial in place, a silent testimony that there *was* a radio in the house.

All the accessories went through a streamlining process. Big wet-storage batteries were replaced by sleek, slim dry cells lined up like wooden soldiers on the floor beneath the set, and they in turn were replaced by handsome rectangular sets, completely self-contained, with dry cells and built-in loudspeakers hidden within the gleaming cabinets.

The streamlined beauties sold for $175, seven weeks' wages for most wage earners. But by that time America was committed to radio for its prime entertainment, and high prices were no longer a deterrent. The big boom had started.

In the beginning, the sets were sold in electric

stores and music shops. Then they began to appear for sale in hardware and automotive stores. Eyebrows were lifted when the radio sets were first offered for sale in drugstores and delicatessens. When the sets sold no matter where displayed, there was a wild scramble of merchants to get in on the flood tide. Nearly every retailer in the country who was not doing well in his regular business tried to sell radios. Plumbers, florists, candy stores, blacksmiths and even undertakers tried to get aboard with a franchise. Most of them made it.

In 1921, the first year after the start of the KDKA broadcasting, the number of radio sets in America doubled from 30,000 to 60,000. But it was in 1922 that the explosion took place. In that year the number of sets increased to 1,500,000, and an RCA official estimated that more than three-quarters of that vast number had been painstakingly constructed by Americans working on kitchen tables and on garage workbenches!

What was feeding the hunger of the millions who clustered around the frail, awkward contraptions?

KDKA started off with the soggy drama of a badly beaten Presidential candidate. The following evening it came on with a potpourri of musical selections.

No one on the small staff was quite sure what to send out over the air and there was much timid experimentation. Music was a solid favorite which could do no wrong. A one-hour program was ample for a dozen selections. To make sure that there was something for everyone, a typical program might present the "Ave Maria," some Strauss waltzes, the "Marseillaise," "Carry Me Back to Old Virginny," a violin solo and, finally, the National Anthem.

It was not the throbbing, full-throated sound of radio music as we know it today. It was thin and far away, a fine thread of sound that seemed to live deep within the cave of the earphones; an elusive sound, one that advanced and retreated whimsically; a harassed sound, one that was threatened almost continuously with faint scratching interference or sudden, surprising pops and crackles. But, with hands pressed tightly to the earphones, brows wrinkled in concentration and all external noises from the kitchen, the living room and the bedrooms quieted, it was an audible and a thrilling sound.

No matter what KDKA presented there was no risk of losing the listening audience. For seven months, from November, 1920, until June 1, 1921, the station was all alone on the air.

Herbert Hoover, still seven years away from the Presidency, spoke in a flat monotone into a microphone on January 15, 1921, appealing for funds for European relief. He set a precedent that politicians all over the country watched with deep interest. Those who listened were not aware that they were the van-

Lee deForest   *U.S. Information Agency*

Professor Reginald Fessenden

Dr. Frank Conrad, world-renowned radio pioneer, at work in his laboratory. *Westinghouse*

The Secretary of the Navy in Washington talking by radio to a ship off the Virginia coast in 1916. *U.S. Navy*

In 1919, this shack on Chicago's North Side housed station 9ZN and a factory for Z-Nith radio products.

Police radios were first used in Europe in Paris, in about 1923.

U.S.I.A.-N.Y. Times

Carl Petersen operating the Byrd Expedition transmitter (also shown at top of facing page) in April, 1930.   *U.S.I.A.*

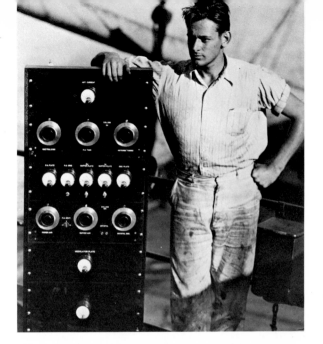

On Admiral Donald B. MacMillan's 1925 expedition to the Arctic, singing eskimos provided dramatic evidence of short wave radio's potential for the world's navies.

Zenith Radio Corporation

In the early 1920's, amateurs constructed big, cumbersome sets.

Everyone listened enthusiastically in the early days of radio.

Westinghouse Photos

KDKA's pioneer broadcast on November 2, 1920, featured the Harding-Cox election returns. The studio's total staff and equipment is shown. *Westinghouse*

The first studio at KDKA was this makeshift tent erected on the top of an eight-story building and used in the summer of 1921. *Westinghouse*

A sound effects engineer on early CBS radio.   *U.S.I.A.-N.Y. Times*

Freeman Gosden (left) and Charles Correll, the popular comedy team of "Amos 'n Andy," in their premiere broadcast in 1928 and on their Sunday evening broadcasts in 1954. *CBS Radio*

WBZ's station and studio occupied somewhat crowded quarters when started by Westinghouse at Springfield, Mass., in 1921.

News commentator Edward R. Murrow in the Ministry of Information, London, during the invasion. *CBS Radio*

The Airborne Command Post of the Strategic Air Command can talk to SAC operations anywhere in the world.

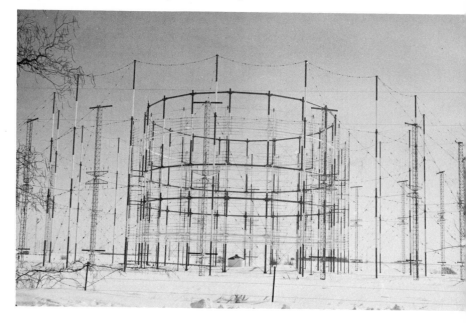

This 172-foot tall Wullenweber antenna at Elkhorn, Nebraska, is one of only three of its kind in the Western world. With other single sideband transmitter and receiver antennas, it has increased the power of SAC's ground-to-air communications to a 45,000-watt capacity.

U.S. Air Force Photo

While one SAC operator logs the transmissions, another passes encoded voice messages to a SAC base on the opposite side of the globe, via the command's single sideband radio network.

U.S. Air Force Photo

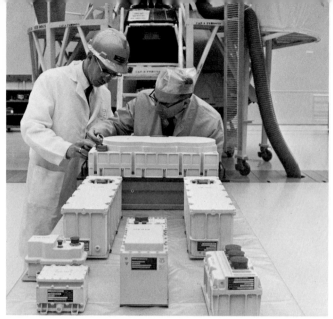

Boxes containing voice portion of the Apollo command module communications system are examined by North American Space Division personnel.

An assembly operator works on Apollo Block II audio center.

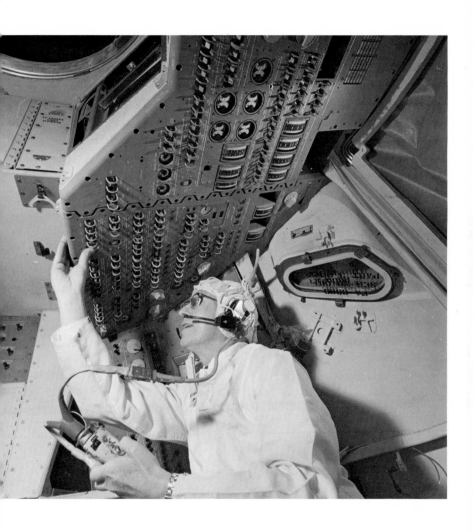

Inside the Apollo command module, a technician tests the astronauts' microphone and voice communication switches on the overhead panel. *North American Aviation, Inc., Skyline Photo*

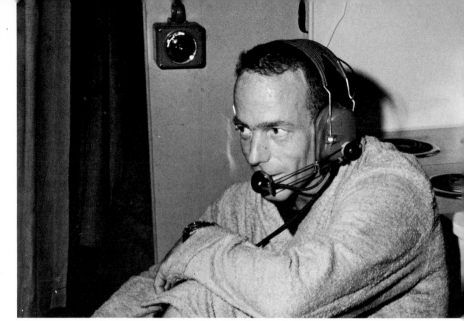

Aquanaut/astronaut M. Scott Carpenter talks from Sealab II, 205 feet below the ocean surface, to astronaut Gordon Cooper during the seventh day of Gemini VI's orbit of the earth.   *Official U.S. Navy Photo*

Messages beamed from the Apollo module after splashdown at sea summon recovery crews.

guard of millions who in the next quarter of a century would be sitting motionless, listening in rapt attention to a succession of voices that would stir the world, move millions to anger, soothe, give hope, scale the Olympian heights of emotion, sound the call to war and announce in grateful tones the return to peace.

A hundred thousand politicians would follow Hoover, their voices trickling out of the earphones and, later, booming out of loudspeakers. Franklin D. Roosevelt, Father Coughlin and Adolph Hitler were among those who waited in the shadows, watching Herbert Hoover's initial attempt to control the fantastic power of radio.

In February, 1921, the station broke out of the confines of the rooftop studio to make the first remote pickup from a downtown Pittsburgh hotel where Colonel Theodore Roosevelt was speaking.

The ability to go far away from the studio, to place a microphone before a willing subject and pipe the results over a telephone line back to the studio for broadcasting opened a wide area of interest. Radio crews from KDKA trudged to theaters to pick up snatches of music and song from performances. They went into hotels for speeches, into churches for religious services, into schools for lectures and into the streets for interviews.

It was an eerie feeling to stand before a microphone and speak. Only the letters that came in from listeners and the quickening sales of radio receiving sets were an indication that a living audience really did exist. But everything the station attempted brought more letters; every new activity was mysteriously translated into more and more sales of radios and radio components.

In April, 1921, still alone on the air, KDKA gave a new twist to programing when it began to broadcast baseball scores, a service that continued throughout the major league season.

For the first six months most of the music going over the air came from records spinning on a turntable in the cramped studio. The success of live broadcasts from several theaters planted the thought that perhaps a live orchestra could be brought into the studio—a less cumbersome method than sending the crews out to theaters laden with bulky equipment.

Westinghouse employees were the musical donors when the attempt was made in mid-May of 1921. The orchestra was placed carefully on the stage of a big auditorium in the Westinghouse plant.

It was a beautiful performance, but it met with nothing but complaining letters from listeners telling how poor it was. The musical notes had boomed and echoed in resonance, utterly destroying the good qualities.

Resonance was something new, and the Westinghouse engineers solved the problem quickly. They moved the entire musical organization to a tent pitched on the roof of the company plant. There, to the accompaniment of mild summer breezes and the occasional hoot of a passing train, live music went out on the air. This time it was a complete success. When the tent finally collapsed in an early fall gale, the lessons of acoustical engineering had been learned, and the new indoor studios, draped with cheap burlap, served their purpose admirably.

Broadcasting farm news each night became a major event at KDKA. At first glance this was not a very striking milestone, but the broadcast bridged the invisible chasm separating rural and urban areas. The two began to mesh.

Stock market reports came on shortly afterwards, building up gradually in response to the stock investment hysteria that was beginning to grip the nation.

Sport events that had once been witnessed only by those who could cram into an arena or into a ball park now were opened up for greater audiences. KDKA covered the Johnny Ray–Johnny Dundee fight in April, 1921, the first of thousands of matches that would follow through the years. Plans were made to cover the Davis Cup tennis matches in the summer of that year, and close on its heels came the first major league baseball broadcast, followed by World Series coverage.

Each event whetted the appetite of those who owned receiving sets, and spurred the interest of investors who wanted to exploit the new field of broadcasting. The Pittsburgh station's monopoly was soon to be shattered.

---
chapter

---
nine

In seven incredible years, the lonely KDKA in Pittsburgh multiplied to 700 stations.

Nine strayed onto the airwaves during 1921. Los Angeles, Chicago, Springfield (Massachusetts), New Haven, Rochelle Park, Newark, New York, Dallas and a sister station in Pittsburgh thrust big transmitting towers into the air and started sending out their versions of market and weather reports, music, concerts and lectures.

Then, in the first four months of 1922, a total of 208 new stations broke into sound. Nearly every state in the Union was represented in the tidal wave of inaugurations. It was like the bursting of a log jam. From then on there was no restraining the rush to get some new call letters on the air.

In the winter of 1921-22, the pattern of living changed in America. Radio was the most important topic—not just of one day, or one week, but every day, every week, every month throughout the long winter.

A San Francisco paper described the discovery that millions of Americans were making. "There is radio music in the air, every night, everywhere. Anybody can hear it at home on a receiving set, which any boy can put together in an hour."

Those fortunate ones who installed the first radio sets on the block were in a magnanimous position, inviting neighbors into their homes to sit in fascinated silence with earphones clamped tight, eyes staring fixedly ahead, mouths open as though to better the concentration.

The broadcasting, in that first year of the new Americana, generally started about five or six o'clock in the evening and lasted for two or three hours.

Those on the East Coast were more fortunate than the pioneer listeners out in the Pacific states. It was early recognized that the darkness of night was kind to the clarity of the signal tones. Shivering addicts stayed with their sets in Philadelphia until eleven o'clock or midnight, achieving the near-miracle of "pulling in" Detroit or—the greatest achievement of all—hearing the thin, reedy voice of the announcer in the Los Angeles radio station.

There was tremendous growth in the number of stations. There was no planning in the phenomenon, no restraint, no restriction. It was one wild electronic party. Stations popped up overnight, the only discernible pattern being the overwhelming desire of the financial backers to get close to the big cities with big listening audiences.

It was a baffling phenomenon. At the time, there was no way in which a radio station investor could get back a return on his investment. There were no commercials as we know them. Westinghouse and General Electric could hope for a profit from the increased sales of radio sets and equipment, but their stations were only a fraction of those coming on the air.

The amateur wireless operators were multiplying at an incredible rate. The big boost in the ranks had come just before America entered World War I, in 1917, when the advent of a cheap Picard crystal detector opened the floodgates. What had been a hobby for bright boys with ample spending money suddenly became available for bright boys who had very little money. The crystal detector did away with the need for expensive vacuum tubes. The tiny bit of silicon, galena or iron pyrites permitted operation every bit as good as the previous tube sets.

World War I brought a temporary halt to the chaotic conditions. Overnight, every amateur and commercial station was closed down under a directive that hinted at the worst possible punishment for any who dared disobey. *Wireless Age* magazine, commenting on the directive that went into effect on April 8, 1917, said, "The President's Executive Order [means] complete disconnection of all pieces of apparatus and antennae, and the sealing and storing of the same. Apparatus which is not dismantled as outlined above is subject to confiscation."

However, the amateurs were to rise again, stronger than before, and oddly enough, they would get their strength from the very ones who sought to "dismantle and store and seal."

The United States Navy sent out a hurried call for wireless operators. Schools were set up in the various Navy districts for preliminary training, and from those smaller schools the candidates moved on to advanced schools of wireless operators. One was at Mare Island, part of a Navy shipyard building facility north of San Francisco; the other was on the campus of Harvard University.

The use of Harvard was not as incongruous as it might sound. The entire university, during World War I, was almost an adjunct of the War Department, fabricating gas masks, giving psychological tests to aviators and decoding German war messages.

For the amateurs it was a fortunate turn of events. When they enlisted with the Navy, they were able to obtain the finest instruction and use the best wireless equipment. They responded in great numbers, and in-

variably the boys who had practiced in their attics led
the classes in advanced radio school. More than 5000
students passed through the "ships," as the buildings
at Harvard were designated, and the old barracks at
Mare Island Navy Yard in California.

When these boys-turned-men returned home after
the war, they waited impatiently for the lifting of the
ban on amateurs. That came in October, 1919. Pad-
locks were removed, old sets were dusted off and the
chattering hordes came back on the air, sometimes a
dozen new stations each night.

In the beginning the amateurs were boosters of
the pioneer station and formed the biggest share of its
unseen audience. But the friendship faded quickly.
When KDKA announced that it was coming on night
after night without cessation, the amateurs rose up in
rebellion. They launched out into the airwaves, ignor-
ing the sopranos who were struggling in the studio
atop the Westinghouse factory. What came out of the
mixup was an odd melee of screeching and horrendous
static.

It was outrageous, but for a time KDKA had to
suffer in silence.

Then the skirmishes were scaled upward and the
amateurs became minor factors in the interference warfare. The commercial stations themselves took on a
favored target, the United States Navy.

As the number of broadcasting stations increased, the interference increased proportionately. In locations along the seacoasts, the high-powered transmitters of the Navy disrupted reception of these stations. At the same time, the broadcast music and speeches from the commercial stations interfered with the flow of official messages on the Navy stations.

The biggest battle of all on the airwaves took place between the commercial stations. They delighted in a form of mutual harassment brought on by a new game, increased power. KDKA had come on the air with a modest 100-watt power; shortly afterward it was increased to 500-watts, which became more or less the rule of thumb for new stations. The power of a station was an indication not only of its strength to send enjoyable programs over a certain distance, but also of its ability to bedevil or completely blank out a lesser power station.

No one quite knew what was happening, yet in the midst of this Mad Hatter's party the radio audience continued to expand explosively. Nothing could dampen the ardor of the swelling millions who were becoming avid listeners of anything that came over the air, from ridiculous descriptions of eggs being fried on hot sidewalks to thrilling blow-by-blow descriptions of world heavyweight title fights.

As yet no way had been devised to make money out of the actual broadcasts.

Staid old American Telephone and Telegraph Company gave the matter much thought, and almost stumbled onto the right answer. Big AT&T decided it could run a radio station, a super radio station, just as it did a telephone exchange. The company spent a large sum of money erecting a high antenna atop a steel building in midtown New York, and beneath the antenna they built a studio crammed with the most elaborate and advanced equipment. It was on the air only a short time when lightning struck the antenna, the studio and the steel building. The steel building had been the cause of intolerable static, so the engineers were happy for an excuse to move to a brick building. There the company continued to explore, in a vague manner, the possibility of getting some return from the free music pouring out over the air.

They broadcast feelers to the unseen audience: "Anyone desiring to use these facilities for radio broadcasting should make arrangements with Mr. A. W. Drake—he can give information relative to periods of information and the charges for the service."

For six months the station, WEAF, received cautious queries. Then, on August 28, 1922, it got a solid bite, the first radio commercial.

The event is of such significance, the forerunner of so much revenue for the industry and so much acrimony for the listeners, that the 10-minute oration deserves to be quoted at length:

MIRACLE OF THE AIR WAVES

This afternoon the radio audience is to be addressed by Mr. Blackwell of the Queensboro Corporation, who through arrangements made by the Griffin Radio Service, Inc., will say a few words concerning Nathaniel Hawthorne and the desirability of fostering the helpful community spirit and the healthful, unconfined home life that were Hawthorne ideals. Ladies and gentlemen: Mr. Blackwell.

It is 58 years since Nathaniel Hawthorne, the greatest of American fictionists, passed away. To honor his memory the Queensboro Corporation, creator and operator of the tenant-owned system of apartment homes at Jackson Heights, New York City, has named its latest group of high-grade dwellings "Hawthorne Court."

I wish to thank those within sound of my voice for the broadcasting opportunity afforded me to urge this vast radio audience to seek the recreation and the daily comfort of the home removed from the congested part of the city, right at the boundaries of God's great outdoors, and within a few minutes by subway from the business section of Manhattan. This sort of residential environment strongly influenced Hawthorne, America's greatest writer of fiction. He analyzed with charming keenness the social spirit of those who had thus happily selected their homes, and he painted the people inhabiting those homes with good-natured relish.

There should be more Hawthorne sermons preached about the utter inadequacy and the general hopelessness of the congested city home. The cry of the heart is for more living room, more chance to unfold, more opportunity to get near to Mother Earth, to play, to romp, to plant and to dig.

So it went, on and on and on, for a full 10 minutes. The cost was $100, about 20 cents a word.

Tidewater Oil Company was next in line, again with a long, tedious oration, although the company no longer has a record of the exact words that were used. On the heels of that came the American Express with an equally long, soporific address.

The reaction was poor. Then a New York clothing house, Browning, King and Company, tried a new soft-sell approach.

"You will now have an hour of dance music by the Browning, King orchestra."

With that, commercials were firmly entrenched, and the eager investors had a solid reason for financing more and more new stations.

chapter
ten

In 1925 the most famous man in America was a radio announcer, Graham McNamee. No other individual could so well illustrate the delirious excitement that swept America, feeding and fattening on the air waves. One man's voice, that of Graham McNamee, poured out an endless torrent of clear crisp words, a new world for millions of citizens.

McNamee was doubly gifted—his radio voice was as clear as a bell, and the intense jump-up-and-down excitement he could personally receive from a prize-fight, a football game, a World Series or a political convention by some magic was captured and projected in the words that gushed from his mouth.

He gave himself entirely to whatever he was watching, and in some matchless manner he was able to carry his audience along with him.

There have been thousands of radio announcers since McNamee's death in 1942 at the age of 53, but none of them ever achieved his prominence. He was the first; he sounded the call that brought America headlong into the age of mass communication. He became the eyes of the land, hurrying from one event to the other. If the event was not startling, his contagious excitement sufficed to give it the needed glow. If the event was exciting, his enthusiasm imparted double and triple dosage to his audience. He could sway millions into a frenzy of excitement.

The vast audiences that McNamee could command with his description of a Saturday afternoon football game, broadcast over New York City station WEAF and its affiliates, should have been a sure harbinger of the wonders yet to come with radio.

But something was basically wrong with the industry.

RCA, a giant in the midst of what should have been unprecedented prosperity, reported a loss to its stockholders.

The rise and fall of radio business enterprises was terrifying. In 1925 there were 258 business firms building radio sets in the United States. Nine years later only five of that number had survived.

The life and death of broadcasting stations was equally perilous. It was estimated that half of the original investors in the new radio stations lost their entire investments within two years' time. But for every one forced into bankruptcy, three others were lined up to take the exciting gamble. Applications to open new stations swamped the Federal authorities. So lax and so vague were the regulatory rules that most applicants, wearied with the delay, went on the air without even the vestige of Federal or State permission.

But while the fascinated investors hurried to toss their money into the electronic well, a sense of uneasiness spread throughout the corporate giants who supposedly were in control of most of the situation. In board rooms of General Electric, Westinghouse and RCA, there were secret conversations, debating the course the companies would take should there be a general collapse of the new radio industry.

There were suggestions that perhaps the radio industry should take strong steps to regulate itself to avoid bringing about its own destruction. But the area was so supercharged, with excitement glossing over the yawning disaster, that the move for self-regulation died quickly. The sprawling industry plunged ahead on its brawling, quarrelsome course.

When Herbert Hoover was Secretary of Commerce he tried to give each applying station a wave length considered adequate for its needs. For a time it appeared as though he had succeeded in calming the turbulent situation—only to have it erupt again, more virulent than ever. A district court in Chicago decided that the Secretary had overstepped his authority, that stations were free to select their own channels, that they could operate them at their free will and that there must be no restraining them.

The entire regulatory system broke down. Confusion reigned. No broadcaster could be sure his speech or music was going out over the airwaves unsullied by whines and snarls and crackles. When the major stations tried to cooperate, utilizing a "gentlemen's agreement" in an effort to protect their million-dollar investments in equipment, they, too, came to unexpected disaster. They scheduled their programs (at first one hour, then expanded to three hours) like careful minuets, one station going off the air just as the other came on. WEAF, the American Telephone and Telegraph Company station in New York, had a gentlemen's agreement with WJZ, the RCA-Westinghouse station in Newark. The agreement didn't prevent WJZ from coming on the air a few minutes early and utterly ruining the finale of a WEAF violin-and-piano program.

With no guidelines, with nothing but a gargantuan appetite for more and more programs, with poor receiving sets that could not easily separate the different channels, compounded by inexperienced engineers who could not control inefficient sending equipment, the wave lengths wandered all about the spectrum, seeking a comfortable place to land.

The squealing, whistling, crackling situation became frantic. Stray wireless messages would *dit-daw* their path into a solemn lecture on beehives; dance music would suddenly override the bellowing tones of an opera singer, competing for the ear of the puzzled audience.

The interference between stations could be summed up in a stepladder of words: annoyance, irritation and nightmare.

By 1927 the confusion had become chaos, and the chaos was leading to utter collapse.

By then there was a listening public of 25,000,000 Americans. There were radio sets in 7,500,000 homes. It took 45 years for the Bell Telephone Company to penetrate that number of homes. It took the utility companies 37 years to install electricity in that number of homes. It took radio just eight years. There had never been a scientific development in the history of the world that was so quickly translated into popular use as was radio broadcasting.

The 25,000,000 listeners were in no mood for the radio industry's collapse. They demanded Federal action.

It had to be Federal action. A radio voice did not move from New York to Chicago alone, or from Philadelphia to Atlanta alone, but simultaneously in every direction. Every syllable uttered into a microphone, every bar of music, every thrilling note of a singer, was a potential destroyer of every other signal. A program beamed from New York out to a waiting audience traveled not only over the darkened city, but all over the United States, and all over the world. The chances for conflicts and disputes were infinite. The troubles were of such magnitude that no mayor of any city, or governor of any state, could cope with them.

The United States Congress alone could provide an answer.

It was a difficult and bitter task for Congress to assume. Every one of the owners of the broadcasting stations, by that time numbering 1000, was a man of influence, very likely wealthy and a campaign contributor back in the home constituency. To move against such bulwarks of local society was asking Congress to cut off its own fingers.

But the anger of the 25,000,000 huddled around the receiving sets in millions of homes scattered around the country finally overcame, partially, the Congressional reluctance.

In 1927 a new statute came on the books. At first it loomed as paragon of severity and justice, in one quick stroke remedying an intolerable condition. A law was passed for the regulation of the radio communication industry that was labeled as "the most severe, the most drastic and the most confining that was ever imposed on any American business."

The first feature of the new law was that 60 days after its passage every radio-transmitting license in effect throughout America was terminated. Boom! Sudden death.

The sword cut neatly through the neck not only of big broadcasting but of everything and everyone capable of transmitting electronic signals. At the time of passage there were nearly 20,000 pieces of transmitting equipment of all kinds in the United States. The new law applied to amateurs; it applied to ships; it applied to the transatlantic and transpacific stations.

When it emerged from the Congress, the law had teeth in it—a penalty of fine or imprisonment for any person who dared operate without a license.

With that out of the way, Congress then set up a new Federal Radio Commission to sit in judgment on the applications for renewal of an old license or the issuance of a new one. Before that was granted, the owner had to demonstrate to the Commission that his station operation would be to the public advantage.

The Commission had a rocky start, with the Senate delaying confirmation of serving members so long that two of the five died before getting full Congressional blessing. The task confronting the members was herculean, and was presented in a sympathetic light by New York City Judge Stephen B. Davis, who gave this commentary one year after the enactment of the Radio Act:

I think it should be said, in answer to much of the criticism that has been directed against the Radio Commission, that the task imposed upon them is no easy one. It is a very simple thing to say to a body of men: "Here are 1000 broadcasting stations in the United States. The physical conditions are such that they cannot all be allowed to exist. It is probably a fact that not more than one-half of them should be allowed to exist.

You therefore will call them in before you. You will separate the sheep from the goats. You will say to a certain number of them, "Enter into the Kingdom," and to the balance, "Get thee into outer darkness," and you will do this although you will realize every time you exclude one you reach into the pocket of its owner and take from him anywhere from a few hundreds to several hundreds of thousands of dollars.

It was asking too much of the Commission to wield such awesome power—and it didn't. Congress had given it the tools to correct the situation, but it hesitated over the impossible task. It did squeeze the 1000 stations down to 700 claimants fighting for 90 available channels.

A lot of abuses were corrected. Interference, if not entirely eliminated, was subdued and made more tolerable.

With all the foot-dragging by the commission, time was gained for the industry to try improved equipment, which widened the broadcasting bands, allowing more and more stations to settle down comfortably in their own tight pastures without having their signals straying over the fences. This was part of an endless search for product improvement that was going on in the research laboratories of the various radio companies.

The people of America settled back in their living rooms, grouped about their Atwater Kent, their Kennedy, their Spartan, Majestic, Zenith, RCA and General Electric radio sets. With the passage of the years the 25,000,000 continued to grow in number, and they sank deeper and deeper into their living room chairs, completely enthralled.

At first just the sound of a voice or the thin, dragging notes of a musical number sufficed. The thrill was in the act of hearing, not in what came over the air. Then the audience became more demanding. Program managers scanned the letters that came to the station. Advertisers were convinced by sacks of mail and reacted.

Suddenly, corner groceries and tubes of toothpaste took on new glamour with the appearance on the air of hard-sell singing and dancing groups burdened with the names of the A&P Gypsies and the Ipana Troubadors. The Gillette Safety Razor got its first lesson in the power of radio advertising, and Maxwell House coffee, with its musical Show Boat, was right behind.

Men's socks received a singing welcome from Billy Jones and Ernie Hare, the Interwoven Pair. Vivien the Coca-Cola girl was a great favorite. And standing in line for early plaudits were other hard-sell groups—the La Palina Smokers, the Cliquot Club Eskimos, the Gold Dust Twins selling soap flakes and the Ever Ready Hour, which plugged dry-cell batteries while presenting good one-act plays.

The same hard sell is apparent in programs today, but not with the same blatant, battering-ram effect. Billy Jones and Ernie Hare, The Interwoven Pair, changed names to fit the product that sponsored them. At times they were known as the Tastee Loafers or the Best Foods Boys.

There were hundreds of other attempts at entertainment. The majority of the singing groups and dance bands soared briefly and died quietly. But when the right formula was found, success was phenomenal.

Graham McName's meteoric rise was paced by other announcers almost equally well known. Major Andrew White for a time threatened to eclipse McNamee, but the same sky-high fame eluded him. Milton Cross got off to a brilliant start and never faltered. Names like Jimmy Wallington and Norman Brokenshire were household words.

Other announcers rode along in fame, sharing the spotlight with the programs they introduced. For more than a decade good announcers were like good jockeys. On the strength of their names alone they guaranteed a large listening audience.

It was the age of Amos 'n' Andy, played by Freeman Gosden and Charles Correll, two vaudeville comedians who started in Chicago radio as "Sam 'n' Henry." For five years the two were solid favorites; then, in a sudden switch of sponsors, they metamorphosed into Amos 'n' Andy. They were on one of the earliest coast-to-coast hookups, and their first programs, much to the chagrin of their toothpaste sponsors, were labeled as terrible flops. Then gradually, very gradually, the new serial took hold. It was the start of one of the most astonishing audience-holding feats ever achieved in radio.

For fifteen minutes each night, five nights a week, the two blackface comedians held most of the nation spellbound. The cast of Negro characters the two invented were woven into the fabric of the times. Amos 'n' Andy introduced an entirely new set of slang expressions of the "I'se regusted" ilk. The comedians kept the nation in a state of suspense with some of the situations they devised. The insistence with which people ran to the radio at the magic moment when the show came on the air was incredible. Even Calvin Coolidge, President of the United States, issued orders that he was not to be disturbed while Andy and the Kingfish, another of the principle characters, were emoting through the radio loudspeakers.

Soap operas had their beginnings in those fascinating times. Phillip Lord created the character of Seth Parker, a good, honest man who invited in his neighbors each Sunday for good, honest hymn singing. The program dripped with emotionalism, and America loved it.

For seven years Seth Parker and his creator, Phillip Lord, rode triumphantly on the airwaves. There was a period of personal disaster for Lord, and then he came roaring back—really roaring—with a new serial, Gangbusters. Again he repeated his original smash success. The hymns and the organ playing gave way to tough cops and the rattle of machine guns.

Entertainers like Bob Hope, Bing Crosby and Jack Benny came on the radio stage and remained for a lifetime.

In 1928 there were already stations that came on the air as early as 6:45 in the morning, and stayed on until midnight.

Within the New York City area alone there were nearly 50 stations in frantic competition to grasp the ear of the unseen audience.

Not everything that came out of the loudspeakers was pure and unsullied. Very soon crooks and speculators learned the wisdom of cultivating the listeners. In that frantic era of stock market speculation, the small handful of years before the crash in 1929, a Chicago investment company sponsored a regular talk on stock market counseling, the broadcaster being an eminent professor of economics at the University of Chicago. Long after the crash of the world-wide House of Insull, which the professor had been touting enthusiastically, it was disclosed that a weekly stipend had been paid by the investment house which was dumping the Insull securities on the unsuspecting public. That in itself was not culpable—but the fact that all of the professor's broadcasting material was written for him in the offices of the investment house was criminal.

The stock market erupted and crashed, and the menacing '30s came over the horizon, followed by the desperate years of the depression. Radio fattened on disaster. When everything else was gone, even the most impoverished could still sit by the blaring instruments and draw a common sustenance with the rest of the nation.

chapter
eleven

Four decades ago, to most Americans, having a radio in the living room was the Mount Everest of accomplishments. It was the focal point of a new way of life. It was the beginning and end of advanced communications. Radio, so American home listeners thought, existed solely for the entertainment of the armchair groups; everything else was unimportant and smacked of gadgetry.

They were wrong.

Living-room radio reached its peak in the '30s, and it stayed there, close to the stars, almost oblivious to other electronic developments that were taking place. These other activities came close to snuffing out the life of living-room radio entirely. They eventually relegated it into a "there-but-not-seen" role, every bit as useful but just as unrecognizable as an electric light bulb.

In radio there are long waves and short waves, the

nomenclature referring to the width of the gaps in the sawtooth pattern of the electronic wave beamed from the sending station. From the beginning of commercial broadcasting at Pittsburgh's station KDKA to this day, the long waves have been the backbone of home broadcasting. These are the waves with which Fessenden first experimented.

The short waves, in the early 1920's, were little understood and much underrated. They were laboratory material, the realm of lone experimenters and amateurs. The short waves were far down on the channel band, safely away from the long-wave commercial excitement, and no one really bothered about them. One historian links the harsh adjective "contempt" with the feeling toward short-wave enthusiasts in the early 1920s.

Dr. Frank Conrad, however, not content with his role in the launching of KDKA on the air, turned his laboratory attention to short waves. He believed they might have some use that could be fitted in with the infant broadcasting efforts. He disbelieved most of the theories then existing about short waves, the most outrageous being that the range of short waves was extremely limited.

It was Conrad who uncovered the intriguing secret of the short waves. Instead of ramming out on a straight line as did the long waves, the short waves tended to skip, bouncing upward to collide with ionized strata above the earth, then bounding earthward again. Conrad likened the action to a stone skipping on a pond. Wherever the wave touched down on earth the signal could be received. And the skipping action was not a random phenomenon. It was almost endless, the signal bouncing and extending until it could travel around the earth in about an eighth of a second.

After Conrad had opened the door, a new army of researchers poured through, performing round-the-clock experimentation with the newly discovered phenomenon. Those early researchers opened up an immense vista. From their efforts were to come radar, static-free FM radio, television and low-power long-distance communications.

Forty years removed from the pioneer research it is easy to list the benefits that came from the initial short-wave efforts, but, as in most great advances, there was at the time much confusion as to the significance of the new knowledge emerging from the laboratories. The land-line telephone people were still hobbled by the thinking that electronic signaling must be viewed in the light of application to private conversations between individuals. The president of the Bell Telephone Company listed reasons why short-wave transmission just wouldn't fit into the communication picture—not only because of its supposed fading tendencies, but because of its nonsecret aspects. The same complaint had been voiced a decade earlier about commercial long-wave radio broadcasting. "Secrecy," the telephone company executive said, "can only be obtained by coding the message in one form or another. And that secrecy could only be maintained against a malicious listener-in only insofar as the integrity of the coding arrangement can be maintained."

The lack of enthusiasm of the telephone companies toward the new short-wave excitement was understandable. The Westinghouse people at KDKA had more than pure scientific reason for developing the new wave form. Using short-wave hookups, Westinghouse hoped to establish chain broadcasting despite the AT&T ban on leased wires for all but AT&T's own chain of stations.

Westinghouse linked its Pittsburgh station with one in Cleveland, Ohio, and the promise was bright. When AT&T withdrew from the radio arena and opened its land lines impartially, for a fee, the need for competing short-wave substitution vanished.

Fortunately the future of short waves did not depend on this one issue, the battle for station-to-station hookups.

Dave Sarnoff of RCA plunged into the new discoveries with his usual enthusiasm. At the very moment when his engineers were telling him in crisp memos that the short waves were not very promising from their standpoint, Sarnoff was mounting speaking platforms to preach the short-wave message. "Very soon," he predicted, "we will signal and talk across both the Atlantic and the Pacific Oceans, using these new short waves rather than the long ones."

Sarnoff's preaching didn't force the issue, but his insistence did send the engineers and scientists back for redoubled efforts. Very soon KDKA broadcasts were being beamed by short wave across the Atlantic and rebroadcast on the British stations. Other equally convincing demonstrations were being made. By 1927 the industry had capitulated, and the short waves were in almost universal use for long-distance communications.

One of the forceful demonstrations came in 1925, with the United States Navy once again the cautious guinea pig.

The Zenith Radio Corporation, in 1925, was just one of the many new radio manufacturers who had sprung up almost overnight to cash in on the insatiable demand for radio sets. The Chicago company was fated, unlike the vast majority of other manufacturers, to survive the mad struggle. One of the reasons for survival was that E. F. McDonald, Jr., the brash young president of Zenith, saw possibilities in short-wave sets that other manufacturers overlooked. As early as 1923 the company had equipped an Arctic exploration vessel with short-wave equipment that performed well, maintaining long-range communications with the States even during the exasperating, endless daylight-hour period, a pitfall for customary long-range communications.

In 1925, radio equipment on naval and merchant vessels depended on long-wave equipment. The systems worked very well during the night, when radio beams were aided by the mysterious ionosphere, the layer of charged particles which completely surrounds the earth. At night these particles form a dense cold layer, acting almost like a radio mirror in the sky, reflecting incoming radio waves over tremendous distances.

But during the daylight hours, the layer warms, the particles thin out and the ionosphere becomes reluctant to "bounce" the radio signals. During the day even the most powerful Navy stations were out of touch with ships and shore stations only a few hundred miles distant.

Wars are fought both day and night, and ships sink just as readily in daylight as in darkness. The situation, for both the United States Navy and the merchant marine, was extremely hazardous.

McDonald, president of the five-year-old Zenith Company, learned that the U.S. Fleet was planning a goodwill tour to New Zealand, Tasmania and Australia. At the same time, the young executive learned, the Arctic explorer Donald MacMillan was heading north to ice fields beyond Baffin Bay in his ship the *Peary*.

McDonald persuaded the Navy to install several short-wave sending and receiving sets on ships heading for the South Pacific, putting them side by side with the heavy, cumbersome, "night-only," long-wave sending apparatus.

Then McDonald personally boarded the Arctic exploration ship, taking with him companion short-wave equipment. The Navy vessels sailed south into the deep Pacific; the *Peary* poked north into the ice fields. When the ships were separated by an incredible 12,000-mile distance, they spoke to each other. There was clear, constant communication in broad daylight. Just to add a convincing touch, McDonald brought a group of Eskimos aboard the *Peary* and had them chant a few songs to the Navy men half a world away.

The clincher for the Navy was when, at the identical time the ships were talking to the Arctic explorers 12,000 miles away, the customary long-wave communications set was unable to raise San Diego, only 8000 miles away!

That was the start of practical use of short-wave radio by the United States Navy. The navies of other nations and the merchant marines of the world soon followed.

The United States Army had the same interest in long-range communications, and while the Army would be reluctant to say that it followed the lead of the Navy in adopting short-wave sets for instant links between far-flung commands, the links were quickly established. But in battlefield communications, the Army followed a strange path before it finally achieved a lightweight, ruggedly reliable voice.

The Motorola Radio Company, another of the young upstarts who fought and survived in the mad melee of the 20s, was destined to play the major role in fulfilling the Army requirements. Motorola, based in Chicago, was not the first company to attempt to build car radios, but it was the first to parlay the experience into a major military contract and a firm footing on the shifting sands of commercial radio survival.

Through most of the 1920s, radio sets still remained comparatively delicate instruments designed to be placed firmly on a table and left alone. They were not meant to be jostled about.

Yet just that requirement had to be met before a car radio would be feasible. The first installation on record took place in 1922. The puzzled owner, a physician, drove his car with the new radio from Missouri to Los Angles without being able to pick up one solitary station. Later it was found that the power plug had been put in backwards.

In 1927 a few sets were built and installed, but the big push did not come until the new decade had rolled around and the nation started the weary trudge through the depression of the 1930s. In 1930 the Galvin Manufacturing Company, casting about frantically for survival, decided to enter the car radio market. The company built most of the 34,000 sets sold that year, watched the number triple and promptly changed the name of the company to the more fitting Motorola. By 1935, nearly 700,000 sets were being sold each year at an average cost of $48.50.

When the Columbia Broadcasting System surveyed the car radio field in that year, they determined that at least 3,000,000 American families had sets installed in autos. The owners' satisfaction was complete. One reported to a CBS query, "I had a puncture in Scranton. About 10 to 15 people gathered around and listened to my radio while the tire was being fixed— women, children and men." Another gave a view of mid-America when he replied, "I'm a traveling salesman—use my car radio continuously, particularly in the evening in Midwestern towns, where sometimes there's not even a movie house for entertainment." Still another salesman reported, "I find it helpful in getting friendly with some of my trade. They come out and I tune in a lively air, and it puts them in a cheerful frame of mind."

They couldn't have always been cheerful. The Motorola engineers had many problems, not only in the basic installation of the sets but also in the product coming from the receiver.

The jolting and pounding the set received traveling over roads that were still in an infantile stage presented an engineering problem of the first magnitude. The packaging problem was formidable, for radios, complete with all their hodgepodge of equipment, were big. Any attempts at miniaturization had been laboratory studies. Overcoming the static noise caused by busy spark plugs spraying an endless stream of interference around the car was another.

In solving these and related problems, Motorola was concerned simply with the task at hand, developing a reliable car radio. Only indirectly was it also getting ready to meet a military need.

The single-pole automobile aerial as we know it today had not yet been invented. Before Motorola engineers could do anything they were forced to find a way to receive incoming signals.

Young engineers sheared the cloth in the top of a car selected for the experiments and exposed the chicken wire hidden beneath the fabric. Two inches of the chicken wire was cut away around the entire roof of the car. What was left was laced back tightly with heavy cord to the frame of the car, leaving an island of wire fully insulated from the car body. The engineers tapped the insulated island, sewed back the cloth and led the wire down to the dashboard.

The experimenters then ripped off the dashboard of the guinea pig automobile, drilled, sawed, bolted and remounted it again as they squeezed the radio components, bigger than a fish-tackle box, into place.

When that was accomplished, they ripped up the floor boards, found a spot between the muffler, the drive shaft and the frame and bolted the cumbersome "B" battery to it. Finally, they hung a speaker near the driver's knee.

They sent the car out for a road test, and it promptly caught fire from its intruding radio installation. The next six experimental installations resulted in the same fiery climax.

When the short circuits and the resulting fires were overcome, another problem asserted itself. The drain of current by the radio was so great that the cars wouldn't start. If the cars did start, they were in imminent danger of "dying" each time the driver tried to pick up speed. The interfering noise coming from the spark plugs often drowned out the sound coming from the speaker. But that problem and the others were gradually whipped. By 1935 Motorola was off and running, selling car radios on a national scale, building a reputation for expertise for its compact, reliable radio sets. Automobile manufacturers, one after another, fell into line. The big "B" battery disappeared, and neat, compact holes for the knobs and dials were thoughtfully drilled and punched by machines at the factory.

In 1936, more than 1,500,000 automobile radios were sold. Compactness and rugged over-the-road operation became an accepted norm for the car radios.

In this day of instant and constant voice communication between the high-flying jets and the ground control stations it is difficult to grasp that it was not until 1929, nine long years after KDKA came on the air in Pittsburgh, that two-way air-ground voice radio communication was established on a commercial airplane.

It was in the early days of United Airlines, when the company was still an awkward group of bits and pieces hastily assembled by financiers. The military, in big, lumbering bombers, had packed aboard the oversize equipment needed for two-way communication, but the luxury of size and weight was denied the shoehorn-tight airplanes that were then making up the commercial fleet. In that year, 1929, the Boeing B-40 carried only four passengers, and the helmeted pilot was still out in the open, buffeted by the slipstream, peering past struts and wings and dangling landing gear while he tried to pick his way over deadly obstacles like the Allegheny Mountains, the Rockies and the Cascades.

Every ounce in the airplane was critical—not only as a test of the motor's ability to lift it aloft over the intervening mountains, but as a possible competitor for the mailbags, the only sure source of revenue.

One-way communication had been tried by some aircraft as early as 1924. The pilot, in scenes reminiscent of Lee deForest in his 1916 experiments, jammed in with tubes, wires and batteries, tried to hear the faint voice coming up from the ground station.

Charles Lindberg, flying in a paper-thin aircraft with even the weight of an extra sandwich a concern, carried no radio, either receiving or sending. He flew in silence across the Atlantic, cut off even from the whisper of sound. All through the '20s this was a condition accepted by pilots. Once they had walked away from the telephone at the home field and climbed into their aircraft, they left the speaking world behind.

Bill Boeing, one of the founders of the line that eventually became United, was talking with his friend Thorp Hiscock about a winter incident in which an aircraft was forced down in a storm. "Everybody knew about that storm but the pilot," Boeing said bitterly. "We've got to find some way to talk to our pilots when they're in the air."

Hiscock, one of the thousands of radio distributors who had bubbled to the surface during the mad '20s, started work on an aircraft radio. Instead of running away from noise and interference, he jammed his experimental radios with all types of interference, subjecting them to the same abuse they would take in a vibrating, bucking airplane.

Hiscock rigged his experimental apparatus on the back of a flatbed truck and drove it over the bumpy dirt roads in the apple orchards of the Yakima Valley. He headed east into the high desert region, snaking into canyons, driving up to mountaintops and slogging along river bottoms, while he tried to keep contact with a sister set he had operating in his Yakima ranch.

Seventeen years before, college student Edwin Armstrong had solved a problem that had completely baffled the early radio researchers. In 1929 Hiscock, a radio distributor and apple grower, solved the problems that had stumped the airline engineers. United adopted his recommendations and started installation on all company planes. From that day on, the commercial pilot was never entirely divorced from the earth.

In 1938 the comedians and the singers and the

bands of renown were shifted slightly from the center of radio popularity. News became a prime topic. The Munich crisis of 1938 was etched into the consciousness of America by the clipped comments of H. V. Kaltenborn during the suspenseful days of the Munich negotiations.

The rush of the real war in September, 1939, carried along with it an entirely new era of broadcasting. Radio became an intimate window on the world, a peep show in which the American audience participated without being hurt. The new voices coming through the loudspeakers—some excited, some rhetorical, others calm and reassuring—came from men who were not only recounting tales of harrowing danger but actually living them.

A young, dark-haired, quietly handsome newspaperman started a lifetime radio and television career when he beamed back to the CBS network the description of a dying city. Eric Sevareid looked from a studio window and spoke in measured tones. "This is Paris at midnight." Like a knowledgeable critic surveying a huge mural, he portrayed the people and the buildings and the stunned incredulity as the city teetered on the brink of collapse.

In the morning Sevareid left the studio, hitched a ride with a friend and then, when the suffocating traffic stopped forward progress, started walking. He turned into the Garden of the Tuileries and sat for a moment staring at an abandoned toy sailboat that drifted in the waters beneath the fountain. He hurried past the Louvre, with its shuttered windows, and then was lost in the crowd. That night, far to the south, he faced a microphone again. Paris had fallen. There was scarcely a ripple of excitement in Sevareid's voice, yet the words, beamed across 3000 miles of the Atlantic, formed a vivid picture for Americans sitting before radio sets throughout the land.

France was overrun, and Sevareid returned to the States briefly before going on to other battlefields, his authoritative voice recounting the shattering progress of a war that finally engulfed even the American listeners.

There was another voice, another city. "This is Edward R. Murrow in London. The German bombers are overhead."

Murrow's terse voice pulled Americans from their chairs until they were all but standing with him on a rooftop in London in the midst of the blitz. They could hear the thudding impact and bursting of bombs as a background for Murrow's voice. He brought the war to millions—not its sensational death and destruction, but a sympathetic portrayal of millions who huddled defiantly in battered tenements or spent long hours in the choking body-smelling odors of jammed subway stations, waiting out the night and the retreat of the bombers.

London survived; the war dragged on. Radio followed the action from one bloody battle to another, from defeat and victory through to the end of the shooting.

chapter twelve

The leather boots of the young military policeman clang on steel mats imbedded in the concrete stairs. His pearl-handled revolver is laced neatly with white cords to a firm position on his hip, but his jaunty blue beret, unlike any other headgear in the entire U.S. Air Force, is perched precariously on his young head. In this place, cut off from the world as most people know it, even the military police are set apart, distinctive.

He leads the way down one flight of stairs, then another and another. An elevator is available, but this clanking approach to the depths lends realism to the entrance.

Out of sight is the Nebraska sky, clear and cloudless over vast stretches of rolling countryside. But deep beneath the sand-colored headquarters building at Offutt Air Force Base, seven miles south of the stockyards and art museums of Omaha, Nebraska, clear skies, dark skies, midnight, noon, today and yesterday are all the same.

The headquarters building at Offutt is an underground fortress. Its power is not in guns or aircraft, or weapons within the fortress, but in the communications network that leaps unseen from this concrete cave to the far corners of the world.

Offutt Air Force Base, a collection of nondescript old buildings and uninspiring new buildings, is a place in the prairie. On the surface, its aging hangars and mathematically correct aircraft runways could be those of any one of a dozen similar U.S. Air Force bases. But beneath the rough grass lawns, it is a different place, a world hidden away from the world, the nerve center of American life-or-death strength in retaliatory nuclear strikes, the headquarters of the Strategic Air Command.

SAC is scarcely 20 years old, and in its short life it has been headed by colorful, powerful four-star flying generals who imbue the Command with a dash, a spirit and an élan that is almost pure Hollywood in its theatrical approach.

Its proponents could argue justly that the approach was needed. The men of SAC have a responsibility never before given to any group of men—the power to destroy the world. They circle in bombers in the sky, waiting, and they sit in 200 underground capsule command posts. At each post are two men who can launch 10 intercontinental ballistic missiles.

Two thousand, three thousand miles away is the President of the United States, the one man who can utter the command "go."

SAC Omaha is the voice bridge between the President and the bombers and between the President and the capsule commanders for the 1000 Minuteman ICBMs.

Perhaps at no other time in history has direct-voice contact between two groups of people, over widely separated distances, been so vital.

The military policeman leads the way along one corridor that bends sharply to the left, then to the right. Overhead, single-eye television cameras peer at those treading the corridor, mute monitors of all who approach. The plain hallway is studded with doors, and behind each are groups of military men with a sprinkling of civilians. Nearly 1000 people work in this three-story underground fortress. If an alarm were to sound, some would evacuate; but there are provisions to keep 700 people alive for two weeks, eating, sleeping, provided with fresh, uncontaminated air while the world overhead reels under the blows and counterblows of nuclear attack.

Visitors pass through one set of doors, then another. Concrete has been poured like a river all about the shell of the fortress. In some places the roof slab is 36 inches thick. Above that is a cushion of earth 46 feet thick. The fortress was designed for survival.

The exact point of control is a war room, half as long as a football field.

In 1955, when the fortress first became operational, the most striking feature of the war room was a series of huge wall maps on which the coming and going of the airborne retaliatory fleet could be plotted. Today the maps are still there, but side by side with them are electronically controlled timing devices and counting devices with far more rigid accuracy—electronic signals that tell the precise time an incoming missile was detected, the probable point of impact, the probable number of seconds until that city-destroying moment.

The retaliatory force of more than 2800 combat aircraft and 1000 Minuteman missiles are tied by invisible lines to the amber lights blinking about the war room.

Spreading out from this fortress within a fortress is a worldwide communication system which has unprecedented demands for instantaneous and unshakable reliability. Neither distance nor atmospheric condition is accepted as an excuse for the operational requirements of the system. When a word is uttered in Omaha, it must be heard around the world.

There is triple safety in the system, a combination of telephones, teletypes and voice radio. It was the need of the Strategic Air Command that led to one of the major advances in post–World War II radio.

SAC's need was satisfied in an Iowa cornfield, only a half-hour's bomber flight from the underground command post.

It is a rousing American success story, featuring the discovery of an electronic genius in the most unlikely place, and with a most unlikely twist. The path of most geniuses leads from the farm to the big city. With Arthur A. Collins, the man destined to forge the final link in voice communications between earth and man-on-the-moon, it was the reverse. Big industry came to him and to the cornfields and took root in the flat farm lands of Iowa.

Collins' father owned tremendous spreads of farmland, and the older man was adamant in the pursuit of mechanization. He made the greatest possible use of scientific aids to farm production.

Young Collins was influenced by his father, but the influence took an unexpected turn. When he was nine years old he had built his first radio set. Later he was riding tractors and wrestling with mechanized farm equipment, but all the time his mind was on voice radio.

In 1924, when he was only 15 years old, he found his shy world shattered when local, then national, newspapers discovered that the farm boy in Cedar Rapids was in regular communication with the explorer Donald MacMillan, at the time far out on the

Arctic ice floes beyond Greenland. Young Collins, sitting in his father's ranch house, was talking excitedly to members of the Arctic expedition, maintaining regular-voice contact at a time when the United States Navy, still bedeviled by the erratic equipment of the early 1920s, had given up in despair.

When he was 21 years old, Collins was back in the newspapers. He was able to talk regularly to Admiral Richard Byrd, who was then leading the first of his three expeditions to the Antarctic regions. When Byrd returned to the United States in June, 1930, he journeyed to Cedar Rapids and sought out the young man who had bridged the tremendous space with his home-made equipment.

The resulting publicity put young Collins firmly into radio manufacturing business. In September, 1933, when all about him business empires were collapsing, Collins diffidently tacked a sign over a basement window and announced to the world the start of the Collins Radio Company. Within 20 years his company was manufacturing nearly two-thirds of all the airborne electronic equipment used throughout the world.

The Collins Radio Company rode a sweeping radio wave of adventure. When Admiral Byrd returned to the Antarctic for his second expedition, starting from Boston on October 11, 1933, equipment from the newly formed company was aboard. With the equipment Byrd planned an ambitious program, one that would lift the world of adventure even closer to armchair travelers. He proposed a regular weekly voice radio broadcast from the Antarctic, to be released directly over the Columbia Broadcasting System network.

The command ship, *Jacob Ruppert,* was pitching and rolling off the coast of Chile when members huddled on the open deck and spoke into microphones. It was a scheduled broadcast; millions of Americans in the States were waiting expectantly by their radios. The seas were so heavy, the rolling of the ship so vicious, that the expedition members had difficulty keeping the proper spacing from the microphone. The Collins radio equipment on board picked up the electronic signals and bounced them to Buenos Aires, whence they were relayed to New York and fed to the waiting audience. The broadcast was a complete success, as were the others that followed regularly from the regions around the South Pole. A new section of the world had been brought into the living rooms of America.

With the coming of World War II, radio was involved completely from the very first flight.

The ferrying of aircraft from the factories of California and Kansas and Missouri and New York into

scores of war theaters throughout the world; the trans-
oceanic paths that were "constructed" over the Atlantic and the Pacific and the Indian Oceans; the voice
control of 1000-unit bomber fleets approaching and
departing from airdromes on missions—all prepared
for the upsurge of postwar civil aircraft activity, a
boom that was intertwined with the reliability of voice
radio communications.

By V-J day in September, 1945, the United States
Army Airways Communications System had built a
network encompassing the entire world. The key was
radio communication, and on that foundation was
built the skein of airways that thread the continents
and the oceans today.

It was in the uneasy postwar period that the
Strategic Air Command moved to the center of the
stage.

From 1945 to 1960, for 15 troubled years, the
Strategic Air Command was responsible for the safety
of the United States, its command exercised in the fleet
of more than 2800 bombers, most of them B-52 giants,
poised at the head of runways at airfields dotted around
the world. This was the famed nuclear deterrent fleet,
the most potent striking force ever assembled. Beginning in 1963, that air fleet was gradually augmented,
then supplanted by a ring of underground silos containing 1000 Minuteman intercontinental ballistic mis-

siles. Each one of these was controlled, ultimately, in the war room of the underground fortress at Offutt Air Force Base in Nebraska.

Today those missiles and giant B-52s still hold joint responsibility for the safety of the United States. And the need for the most sophisticated voice radio control is still paramount. It was out of the needs of the SAC central command that voice radio made one of its most dramatic leaps.

With a flick of a switch in the underground fortress, the group of officers in Nebraska can be in instant contact with a SAC communication base in Greenland. With another effortless touch, the voice of the duty officer at a SAC bomber base in Guam speaks out; another switch, and the voice of an officer on duty underground with the Minuteman missiles hidden in the state of Montana comes into the room. Spain reports, Alaska, a SAC base in Florida, then the Philippines, and so on around the world.

Until 1956 there was one weak link in the intricate system of SAC world-wide communications. The mysterious auroral absorption zones over the North Pole, a vital operating area for SAC, still dragged like a shifting curtain, a constant threat to the sure maintenance of communications between a patrolling bomber and the SAC headquarters in Omaha.

Art Collins started working on the weak link as

early as 1949. His goal was a voice radio system that would permit instant contact from the Omaha headquarters with a U.S. Strategic Air Command plane flying anywhere in the world. He finally had the system —compact, clear and with an amazing capacity. To prove that the new principle would work, Collins went aboard a SAC bomber, flew over the North Pole and talked simultaneously with another bomber flying over the South Pole, half a world away.

Single-side band radio transmission, the new marvel that Collins introduced, was a great step forward, a neat slicing of the spectrum of radio frequencies, a seemingly simple move that vastly increased the number of airborne radio conversations that could be carried on. It was a key factor that carried SAC and the scientific world into a new era of communications.

But while the advances on the military side were in giant strides, radio in the living room for a brief postwar period seemed faced with extinction.

The advent of television after World War II brought gloomy predictions that home radio sets were doomed. Yet the sale of radio sets continued to boom, and each year a new record was set. Today the miracle of clear, lifelike tones coming out of instruments, some of them as small as the palm of one's hand, is taken for granted. Three, four, five radios in each home are accepted. There are 560,000,000 radio sets in use throughout the world.

Radio in the living room, radio in automobiles, radio in world-girdling airplanes— the human voice had only one place to go, across the bridge of space, outward to the moon.

chapter

thirteen

With a few notable exceptions the Atlantic coast of Florida has undergone few changes since the beginning of time. The hard-packed sand of the beaches, the relentless, lulling action of the lukewarm waves, the hot blue skies crowded with puffed, humid storm clouds—all these are the same.

The changes visible on the surface of the water have been gradual. More shrimp boats amble each evening out to the fishing grounds off Cape Kennedy, then turn and amble home again against the flaming red of the new sun each morning. An occasional nuclear submarine pokes to the surface and threads a cautious path through shallow waters to a sheltered dock at Port Canaveral for the on-loading of Polaris missiles to be test-fired down the South Atlantic.

On the near-ocean land, in a short, crowded stretch ripped from mangrove swamps, there have been changes. More cars are jammed on the strip of macadam on the highway several hundred yards inland from the beach; more motels are filling up the sand between the road and the beach; and more night lights are apparent at the world-renowned Kennedy Space Center, the American spaceport, the leap-off site for the journey to the moon.

The spaceport is a collection of red metal towers and pure white buildings raw in the glaring sun. The buildings and the launch towers vary in size from the big to the gigantic. At night a soothing darkness hides the rawness and the stark angularity of the towers, and only the gleaming lights betray the unending effort to complete the first portion of the greatest epic of all time, the dispatch of men to the moon and their safe return home.

For 13 years, the hook of sand at Cape Kennedy, once a deadly menace for Spanish galleons, has had a mesmerizing effect on increasing numbers of Americans. The first big rocket went aloft from the mangrove and muck in 1953, but it wasn't until 1957, with the launching of the first Thor, an intermediate-range ballistic missile, that the Cape started to send out the pulsing radiations that were to draw hordes of spectators. In the pre-jet days, the hurried flight over from Los Angeles to watch a missile launch in Florida was a 12-hour horror culminating in the early morning hours at Cocoa Beach motels that have long since disappeared or are now slowly disintegrating in green corrosion and mildew.

Despite the hurry, hurry, hurry over from the Coast, inevitably there was a wait of one day, two days, three days, for the mysteries of the early rocket systems to unravel themselves and consent to be blasted aloft.

In those days, almost prehistoric now, the tough, chunky Redstone rocket, a military construction capable of flinging nuclear warheads 200 miles toward enemy entrenchments, was the favorite. It was fat, slow, impressively reliable, lifting off from the pad ponderously like a rounded moving van climbing into the sky.

But its ponderous reliability brought fame. It was the vehicle chosen for the first inching step by man into space. It happened on May 5, 1961, when Alan Shepard climbed inside the shoehorn-tight Mercury capsule atop the blunt nose of the Redstone, and gave himself over to God, his country and the still far-future Apollo moon program.

He had been awakened at 1 A.M., been examined, fed carefully, then suited up and brought out to the site. He went inside the elevator and went up, up, up to the 70-foot level in the shaft of open steel, and then wedged himself into the capsule.

There he was strapped, helmeted and patted comfortingly on the shoulder. Then the attendant withdrew.

There were holds, more holds, and then all was ready. More than a mile away, on an old wooden platform, hundreds of newsmen, all of them tired, most of them uneasy, a few irritable from an excess of hilarity of the night before, waited impatiently. Those who stood by open telephone lines leading back to distant newspaper offices were the most impatient of all. A few, bored beyond relief, played cards on the bare wooden platform, shivering as the new sun came up and played hide-and-seek through the broken clouds.

Then the familiar countdown, the callout of an orderly procession of mechanical events, was begun. The voice from the loudspeaker crackling over the heads of the reporters called off the final minute in 10-second intervals. The count was carried down, 9 seconds, 8, 7, 6. . . . At zero there was an uneasy silence. The automatic launch sequence had taken over. Deep within the missile a liquid-oxygen vent valve slammed shut.

At 9:34 A.M. there was a bright flash of flame, a rumble, then a thunderous roar. The needle nose of the missile stirred almost imperceptibly. Then, in a moment of magic, it inched upward.

From the pad came the cry that was afterward to be an embarrassment to the veteran observers: "Go, baby, go!" From others, more reserved, came a long, grunting approval, as though the observers were trying to develop a deep-throated resonance that would harmonize with the slender stick of metal climbing high on a pillar flame into the sky. They knew the man atop the blunt vehicle was facing death that could come in a bright ball of flame. They grunted half in expectation, half in prayerful sympathy, that he would survive the ordeal.

The upward climb became a glide, a run, a rush. The noise escaping over the flat Florida landscape, was one vast continuous belch of sound.

Thanks to Marconi and deForest and Fessenden and the long line of unsung heroes who had prepared the way, Shepard was not entirely alone.

Two and three-tenths seconds after lift-off, as the Redstone generated its full 83,000 pounds of thrust, the astronaut spoke into the tiny stick microphone poised before his lips. His voice, rasping and strained, came back across the widening river of space to his eager listeners, "Ahh, roger, lift-off and the clock is started. Yes, sir, reading you loud and clear. This is Freedom 7; the fuel is go; I'm pulling 1.2 g; the cabin is at $14\frac{1}{2}$ psi; the oxygen is go. Freedom 7 is still go."

Freedom 7 went all the way. For 15 minutes Shepard soared in a majestic ballistic arc at 5000 miles per hour, until he plummeted down into the ocean

302 miles southeast of Cape Kennedy (then known as Cape Canaveral).

Nine months later John Glenn became the first American to go into orbit around the earth. Forty million people in this country listened by their radios and television sets as he spoke from his speeding spacecraft. Other astronauts followed. Walter Schirra, in one of his earth orbits, gave a hint of the wonders to come when he spoke to the listening world, "I'm looking at the United States. The spacecraft is starting to pitch up slightly. Now I can see the moon, which I'm sure no one in the States can see as well as I can right now."

The stage was set for Apollo.

The dispatch of three men to the moon, and their safe return to earth, is a project of such magnitude that it almost swamps the imagination. One of the leading officials of the National Aeronautics and Space Administration said of the project, "No one underestimates the enormity of the task that has been accomplished and that is still awaiting us. All the blood, sweat and anguish that went into the building of the Pyramids, all the hurry-up scientific genius that went into the Manhattan atomic bomb development project, all the 60-odd years of preparatory aircraft work that went into the development of the supersonic transport—

those three combined just begin to equal what Americans have accomplished in the past eight years."

In a project as critical as the Apollo, where literally hundreds and thousands of systems and subsystems must work perfectly in order to achieve success for the venture, it is foolish to single out any one system as being more vital than the others. Voice radio is critical, but so are the engines, the guidance and navigation systems, and so are the systems that maintain an even, acceptable living atmosphere within the tight metal world hurtling to the moon.

The fascination of voice radio comes in its human quality. A multimillion-pound thrust cluster of rocket engines is a scientific masterpiece, but it lacks the intimate excitement of a man's voice crackling across a hundred thousand miles of space. The guidance system of the moon vehicle is a miracle in itself, but that miracle is lost in the astronaut's voice reporting calmly, "We're making our descent to the surface of the moon."

Voice radio is the link between the men in space and the men on earth who will be monitoring the expedition. But the voice will have a far more exciting impact, for the astronauts' words will be heard over a world-wide radio network. Everything that is said and done will be brought into the living rooms of the world. The fascinations for the armchair listeners will not be entirely with the scientific achievement.

From earliest time, exploration and threat of personal disaster have gone hand in hand. Men push into the unknown, never quite certain what will be just beyond the ocean's horizon, around the next jungle trail or in the next swelling dip of a western prairie.

Columbus and his men were fearful of death and survived; Magellan, with more bravado, was prepared for death on his round-the-world voyage and met it; Livingstone, in the heart of Africa, walked with the knowledge that his disappearance from the civilized world could change quickly into permanent disaster. Had voice radio been available in those times, every forward movement into danger would have been followed by a tense, expectant audience.

When the three astronauts duck downward, wiggle through the hatch door opening and strap themselves on the couches inside the command module, awaiting the momentous start of a 240,000 mile journey, they will be participating in a voyage where death will be a constant hazard for every split second of the entire 14-day journey.

The pressurized world in which they live, a metal room scarcely 12 feet in diameter at its widest point, scarcely man-high in the center, is cut off entirely from the world they have left and from the moon world to which they are journeying. At the very outset, the three men are exposed to the possibility of one of the most devastating man-caused, non-nuclear explosions ever witnessed, a fireball that would consume 3000 tons of highly volatile fuel.

The astronauts are volunteers. They know the danger that awaits them, not only in the adventure itself, but in the testing period that precedes the actual flight. On January 27, 1967, Virgil Grissom and Edward White, both veterans of Earth orbital flights, and Roger Chaffee, all met a fiery death while participating in a simulated countdown during the test of an Apollo moon spacecraft at NASA's Kennedy Space Center in Florida.

The hazard will exist from the very moment the men enter the module, during the hours of preparatory countdown, through the perilous seconds of lift-off and through eight-and-a-half minutes of boosted flight.

Should the entire towering stack erupt in explosive flames, a safety tower will explode in its own fury and rip away the module containing the three astronauts, hurtling them to a jarring landing in the ocean not far away. It would be a harrowing experience, but the three would survive.

Should the explosion take place in the first two and a half minutes of flight, with the space vehicle 38 miles in the air, and ramming forward at 5360 miles an hour, then the module would still be ripped away, and the men would still fall to waiting picket boats on the heaving surface of the Atlantic.

In all cases the three men would remain inside the module until it landed on the surface of the sea. If an unprotected man were to try to eject himself from a command module at any altitude 10 miles or more above the earth, he would die. At 50,000 feet and higher, the human blood boils within the veins, and death is instantaneous.

The imminence of violent death, swift and uncertain, will exist through every split second of the voyage. It is as though Livingstone had to make his jungle trek down a trail solid with poisonous snakes, as though Columbus had set off to swim through shark-infested waters from Spain to the unknown America.

Should the two men who drop down to the surface of the moon be unable to start the engine that will thrust them upward again to the command module circling in moon orbit overhead, they will be doomed. The entire world will listen as they recount their activities in the four or five days they will remain alive until the depletion of their oxygen supply.

The same harrowing experience will await the astronauts and the listening world should the command module miscalculate the minuscule re-entry corridor when it returns from the moon and attempts the perilous entry into earth's atmosphere. Should that happen, the vessel would bounce outward, circling and circling, unable to penetrate quickly enough to keep the men alive before the 14-day safety margin is expended.

Granted the success of the venture, granted the smooth synchronization of one exquisite detail after the other, granted the mastery of the most complex navigation maneuvers ever asked of mortals, the listening world will be subjected to one final drama, this one with every ear straining into a noiseless void.

The command module, as it drops back to earth, will enter a 16-minute period of metal scorching heat. The three men will be protected within a glowing module whose temperatures reach peaks nearly 10 times higher than inside a kitchen oven. For four minutes of that fiery hell, the astronauts, for the only time during the 14-day voyage, will be completely mute, their voice radio systems the victims of an unsolved phenomenon that blacks out all radio communications during the hottest moments of re-entry.

Untold millions throughout the world will be waiting, tense and uncertain, until once again they hear the scratchy voices come from the command module. Then the craft will plummet into the sea, and the voyage will be ended.

These are the grim possibilities. But 30 billion dollars and eight years of effort have been spent to preclude any disaster.

In the command module the voice radio is entwined with the data system, a marriage that would have completely befuddled Reginald Fessenden, and still puzzles most laymen. Streaming from the command module back to earth are broad bands of waves on which the voice and the entire spread of telemetry information are mingled. At the same time an astronaut is reporting that he has just completed his sleep cycle, taken his blood pressure and assumed his monitoring of the guidance and control system, a cacophony of pips and bleeps will be giving the position of the spacecraft, the temperature inside the module, the amount of fuel left in the propulsion engine tanks and the microseconds remaining before a course correction will be necessary.

The Collins Radio Company supplies the communications and data system for the venture as subcontractor to the North American Rockwell Corporation, builders of the Apollo command and service modules in Downey, California.

The system is the only link between the spacecraft and earth. Through it pass all voice communications, as well as tracking and ranging information, telemetry, television and recovery beacon signals.

When an astronaut stretches out full length on the take-off couch, he puts on a radio headset, the most conspicuous part of which is a long thin tube which sweeps around his cheek and poses a blunt end before his lips. The words of each astronaut are piped through the matchstick-size tube in front of his mouth to a microphone which is built into the ear assembly. Each crewman wears two of the tube-mike combinations. If one fails, he still has the other in reserve.

The same safety features of double units are in the earphones. There are two, each entirely independent of the other. Each acts as a backup for the other in case of failure.

As the astronaut stretches out in the prone position for take-off and landing, immediately over his head is a panel of dials and switches. An umbilical cord stretches from his headset to the panel, his personal audio control room.

Each astronaut uses his headset for all voice communications. He controls what comes into the headset and where he will send his voice by means of the audio control panel within arm's reach overhead. All three astronauts have identical panels.

Transmission is initiated or stopped either by the push-to-talk switches in the astronaut's communications control head or by a voice-operated relay. In an emergency, the push-to-talk switches can be used like telegraph keys to tap out messages in Morse code—a 60-year reversal to the early days of Fessenden, deForest and Armstrong.

The complexity and the range of communications,

however, have gone beyond anything ever contemplated by the three pioneers.

The whole scope of radio knowledge has been embraced—high frequency, very high frequency, FM and AM and, the most striking change of all, unified S-band equipment. The jargon of radio codes and symbols is endless. Some types of equipment are used solely during the preliminary orbits about the earth while the astronauts are testing equipment and jockeying for the crucial outward push that will tear the module from the clutches of earth's gravity and send it hurtling outward at 25,000 miles an hour toward the moon. In that tense period, when momentous decisions are being made, the voices of the astronauts will go from the earth-circling module down to one in the chain of receiving stations set up around the world by the National Aeronautics and Space Administration, then will be relayed back to the master control center in Houston, Texas.

When the decision has been made and the craft has vaulted out to deep space, feeling for the imperceptible pull of moon gravity, communications will be on the new S-band.

The man who first steps out of the lunar module onto the surface of the moon will have as his initial chore the erection of a retractable antenna. When he starts to walk away from the grounded module, he will speak into his microphone, and his voice will go on electronic beams from his headset to the communication panel within the grounded lunar module. From there the beam will be sent over the newly erected antenna directly to Houston.

In 1962, when the design of the moon module was frozen, so that full speed ahead could be given to the historic moon voyage, all design efforts in the voice radio equipment in the Apollo module were likewise frozen.

As a result, the state of the art has gone far beyond the equipment now being used for the moon voyage. The bouncing of radio waves and television signals from man-made satellites is now commonplace. Soon the world will be tied in a net of combined radio and land telephone lines that will make home dialing to any point on earth a distinct feasibility. The world of Fessenden, deForest and Armstrong has not stood still waiting for the moon conquest.

But nothing will take away from the electric excitement of that coming moment when man-on-the-moon talks to man-on-earth.

On that day the entire world will be listening.

On that day the universe will begin to talk.

Bibliography

Archer, G. L., HISTORY OF RADIO, The American Historical Society, Inc., New York, 1938

Carneal, Georgette, CONQUEROR OF SPACE, Liveright Publishing Corporation, New York, 1930

Chase, Francis, Jr., SOUND AND FURY, Harper & Brothers, New York and London, 1942

Coe, Douglas, MARCONI, PIONEER OF RADIO, Julian Messner, Inc., New York, 1943

Dunlap, Orrin E., Jr., COMMUNICATIONS IN SPACE, Harper & Row, New York and London, 1962

Fessenden, Helen M., FESSENDEN: BUILDER OF TOMORROW, Coward-McCann Inc., New York, 1940

Howeth, L. S., Captain, USN, HISTORY OF COMMUNICATIONS—ELECTRONICS IN THE UNITED STATES NAVY, Govt. Printing Office

Lessing, Lawrence, MAN OF HIGH FIDELITY: EDWIN HOWARD ARMSTRONG, L. B. Lippincott Co., Philadelphia, 1956

Lyons, Eugene, DAVID SARNOFF, Harper and Row, New York, 1966

Maclaurin, Rupert W., with the technical assistance of Joyce, Harman R., INVENTION AND INNOVATION IN THE RADIO INDUSTRY, The Macmillan Co., New York, 1949

Slate, Sam J., and Cook, Joe, IT SOUNDS IMPOSSIBLE, The Macmillan Co., New York, 1963

Index

INDEX

Bronx, The, 91
Brooklyn Navy Yard, 62
Buenos Aires, 162
Byrd, Admiral Richard, 161-162

California, 162
Canada, 40, 106
Cape Canaveral, 172
Cape Cod, 52
Cape Hatteras, 44
Cape Henry, 44
Cape Horn, 55
Cape Kennedy, 167-168, 172
Caruso, Enrico, 62, 91
Cascades, 151
Cedar Rapids, Iowa, 160
Chaffee, Roger, 175
Chesapeake Bay, 32, 47, 57
Chicago, 18-21, 117, 128, 130, 143, 146
Chicago, University of, 138
Chile, 60
China, 24
Cleveland, Ohio, 142
Cliquot Club Eskimos, 135
Coast Guard, 68
Cobb Island, 43
Cocoa Beach, 169
Colliers magazine, 57
Collins, Arthur A., 160-162, 165
Collins Radio Co., 161, 178
Colorado, 59-60
Columbia Broadcasting System, 147, 162
Columbia University, 65, 69, 73-74
Columbus, Christopher, 174, 176

Commercial Cable Co., 77-78
Congress of the United States, 131-133
Connecticut, 52-54, 56
Conrad, Dr. Frank, 94-99, 103-104, 140-141
Coolidge, Pres. Calvin, 136
Correll, Charles, 136
Costa Rica, 25
Coughlin, Father, 113
Cox, James, 101, 103
Crookes, William, 10, 12, 15
Crosby, Bing, 137
Cross, Milton, 135
Cuba, 25

Dallas, Texas, 117
Davis Cup, 115
Davis, H. P., 98-102, 107
Davis, Judge Stephen B., 132
deForest, Lee, 17-28, 44-45, 47-64, 71-77, 88, 91, 94, 96, 98, 151, 171, 179, 181
deForest Radio Telephone Co., 63-64
Detroit, Mich., 118
DeVeaux Military College, 40
Downey, Calif., 178
Drake, A. W., 123
Dundee, Johnny, 115
Duryea brothers, 107

East Bolton, 40
Edison, Thomas A., 30, 40-41, 46, 71-72
Eiffel Tower, 61, 85
England, 13, 16, 19, 40, 80, 88-89

Index

187

INDEX

INDEX

Ray, Johnny, 115
Redstone rocket, 169-171
regenerative feedback circuit, 73
Rhode Island, 21, 53
Rhode Island Sound, 54
Richmond, 48
Rio de Janeiro, 57
Roanoke Island, 44
Rochelle Park, 117
Rocky Mountains, 151
Roosevelt, Franklin D., 113
Roosevelt, Pres. Theodore, 55-56, 113
Rosenberg, L. H., 102
Rotterdam, 7

San Diego, Calif., 145
San Francisco, 24, 61, 71, 84, 120
San Francisco Bay, 84
Sarnoff, David, 77-83, 89-90, 142
Schenectary, 29, 31, 35
Schirra, Walter, 172
Scotland, 34-35
Senate of the United States, 132
Sevareid, Eric, 153
Shepard, Alan, 169-171
Show Boat, 134
Signal Corps, U.S. Army, 93-94
Sino–Russian War, 24
South Pole, 162-165
Spain, 164, 176
Spartan, 134
Springfield, Mass., 30, 117
Straits of Magellan, 60
Strategic Air Command, 157-159, 163-165
superheterodyne circuit, 110

Supreme Court, U.S., 76

Taft, Pres. William, 81
Tasmania, 145
Tastee Loafers, 135
Thomas, William, 102
Thor missile, 168
Titanic, 79-81
Towers, Walter Kellogg, 86
Trinity Church, 93
Tuileries, 154

United Airlines, 150-151
United Fruit Company, 25, 32, 88, 90
United States of America, 12, 18, 21, 25-26, 55, 62-63, 77, 80, 82, 87, 89-90, 106, 127, 130, 132-133, 163, 172

Vail, Theodore, 84
Virginia, 52-53
Vivien, the Coca-Cola girl, 135

Wall Street, 23
Wallington, Jimmy, 135
Wanamaker Department Store, 80
War Department, U.S., 120
Washington, D.C., 43, 48, 49, 52, 55, 93, 95
WEAF, 123, 127, 129
Weather Bureau U.S., 42-44
Welsbach gas mantle, 72
West Selene, 7, 38
Western Electric Co., 19-20
Western Union, 12, 26

Index

About the Author

EDWARD A. HERRON was born June 5, 1912 in Philadelphia, Pennsylvania. After graduation from St. Joseph's College, he shipped out as a merchant seaman. Then with royalties from a book published during undergraduate days, he headed for Alaska where he worked in the gold mines. During this period he started selling short stories and articles, mostly dealing with Alaska. Mr. Herron and his wife, with their children who were born in Alaska, now live in California, where he is a writer in the space program. It was while preparing a series of articles on the Apollo moon module that he became interested in the vast strides that had been made in the area of voice radio communications. He wrote this book to give credit to the scores of radio pioneers upon whom the safety of today's astronauts depends, and to link those pioneers with the exciting events of the space age.